BEGINNINGS OF THE AMERICAN
RECTANGULAR LAND SURVEY SYSTEM

This is a volume in the Arno Press collection

THE MANAGEMENT OF PUBLIC LANDS IN THE UNITED STATES

Advisory Editor
Stuart Bruchey

Research Associate
Eleanor Bruchey

Editorial Board
Marion Clawson
Paul W. Gates

*See last pages of this volume
for a complete list of titles*

BEGINNINGS OF THE AMERICAN RECTANGULAR LAND SURVEY SYSTEM, 1784-1800

WILLIAM D. PATTISON

ARNO PRESS
A New York Times Company
New York • 1979

Editorial Supervision: JOSEPH CELLINI

Reprint Edition 1979 by Arno Press Inc.

Copyright © 1957 by William D. Pattison

Reprinted by permission of the Ohio
Historical Society

THE MANAGEMENT OF PUBLIC LANDS IN THE UNITED STATES
ISBN for complete set: 0-405-11315-3
See last pages of this volume for titles.

Manufactured in the United States of America

Library of Congress Cataloging in Publication Data

Pattison, William David, 1921-
  Beginnings of the American rectangular land survey.

  (The Management of public lands in the United States.
  Reprint of the ed. published by the Ohio Historical
Society, Columbus.
  Bibliography: p.
  1. Surveying--United States--History. 2. Surveying--
Public lands--United States--History. I. Title.
II. Series.
[TA521.P37 1979]   526.9'2'0973   78-56654
ISBN 0-405-11350-1

BEGINNINGS OF THE
AMERICAN RECTANGULAR LAND SURVEY
SYSTEM, 1784-1800

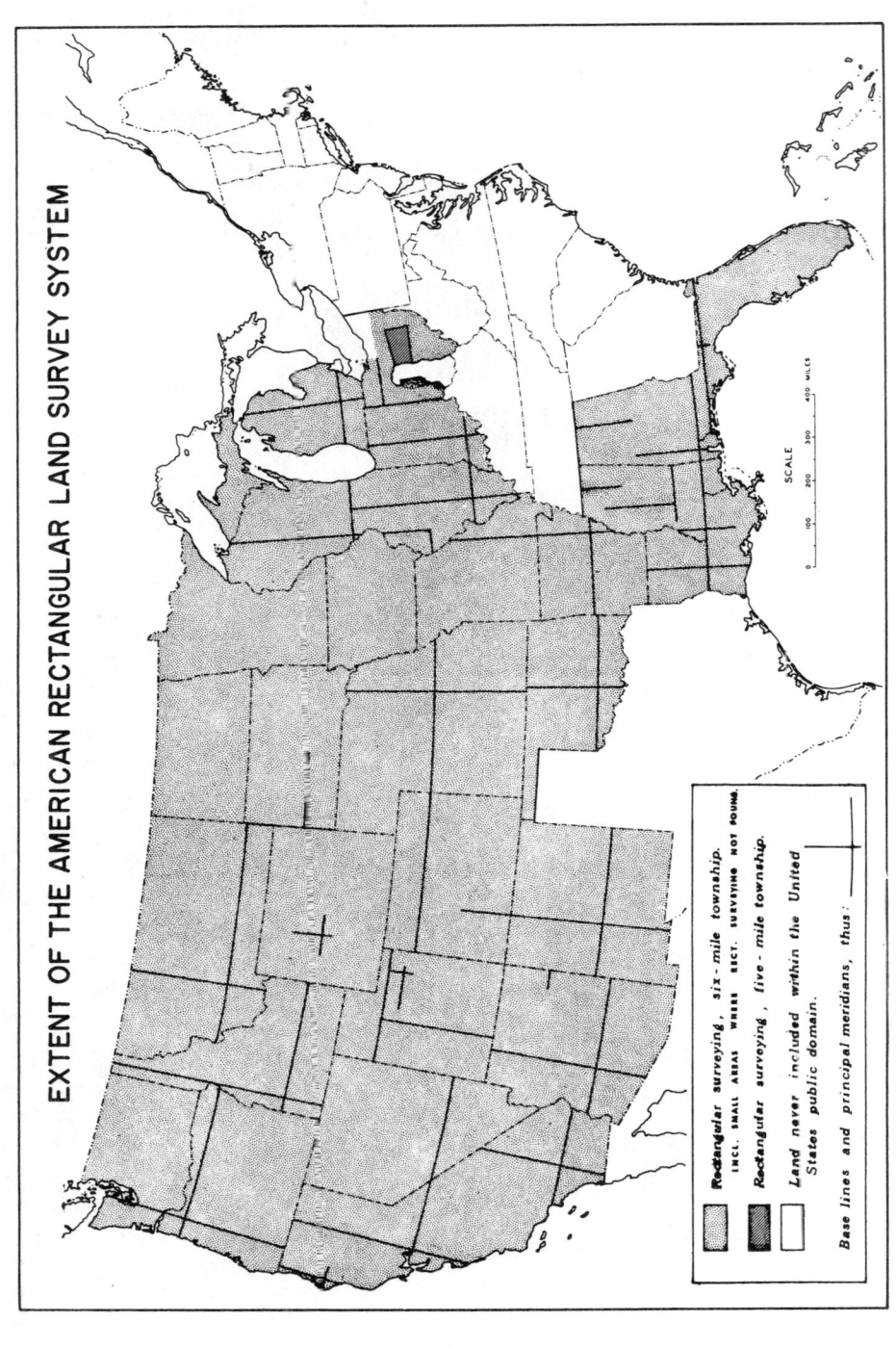

BEGINNINGS OF THE AMERICAN RECTANGULAR LAND SURVEY SYSTEM, 1784-1800

WILLIAM D. PATTISON

THE OHIO HISTORICAL SOCIETY

COLUMBUS, OHIO

Copyright by William D. Pattison. All rights reserved. Published 1957. Second printing 1964. Published originally as University of Chicago Department of Geography Research Paper 50, by the University of Chicago Press. All rights to publication acquired by The Ohio Historical Society, 1966.

The Ohio Historical Society - 1970

PREFACE

In a sense, this study began in London, England, nearly five years ago, when my attention was drawn to the United States public land surveys by H. C. Darby of the Department of Geography, University College London. Interest centered at first in finding out uses to which the descriptive content of the public land survey records had been put, and I undertook an inquiry along this line which was later completed at the Department of Geography, Indiana University, under the sponsorship of Norman J. G. Pounds. Meanwhile, the public land survey system as a whole had become increasingly interesting, and preliminary research in Washington and elsewhere had suggested to me opportunities for contributing to an understanding of the system. It was as a returned student in the Department of Geography, University of Chicago, that I fixed upon the scope of the present study, in September, 1955, since which time research and writing have gone forward under the supervision of Wesley C. Calef.

For assistance in research my thanks are extended primarily to Herman R. Friis and other members of the staff of the National Archives. For points of advice I wish to thank Earl G. Harrington of the Bureau of Land Management and F. J. Marschner of the Department of Agriculture. Further thanks are owed, for the use of unpublished material, to institutions named in the bibliography at the end of the study. For published material, I am indebted to the Library of Congress and to Harper Memorial Library, University of Chicago.

Two persons share credit for completion of the study at the present time. They are Elizabeth Ross and Joan Davis, who, in Washington and Chicago, respectively, held forth a helping hand in time of need.

August, 1957
Chicago, Illinois   William D. Pattison

TABLE OF CONTENTS

	Page
PREFACE	iii
LIST OF ILLUSTRATIONS	vii
ERRATA	viii
INTRODUCTION	1

Chapter

PART I: THE BASIS IN LAW

I. THREE PRIOR CONSIDERATIONS 3

    State Cessions
    Indian Cessions
    Jefferson's Plan for Western States

II. THE ORIGINAL PLAN FOR A FEDERAL RECTANGULAR LAND SURVEY 37

    The Southern System Reformed
    Division into Hundreds
    The Geographical Mile
    Rectangles and Meridians
    The Square-Mile Section
    The Question of Origins
    Review

III. ADDITIONAL SURVEYING PROVISIONS IN THE FIRST PROPOSED NATIONAL LAND ORDINANCE 68

    Registers and Surveyors
    Trees, Chain, Plat and Compass
    The Identification of Lots
    Review

IV. THE LAND ORDINANCE OF 1785 82

    The Principle of Prior Survey
    The Rectangular Grid Retained
    Surveying and Numbering
    Geographer and Surveyors
    The Place of Beginning
    Review

PART II: SURVEY OF THE SEVEN RANGES

V. THE FIRST SCENE OF SURVEY 105

    Pittsburgh and the Road from the East
    Fort McIntosh
    Settlements on the Upper Ohio River
    Indian Tribes
    Lay of the Land in the Seven Ranges

VI. A CHRONICLE OF SURVEYING, 1785-1788 119

    Establishment of the Beginning Point
    The Geographer and Surveyors Assemble

Chapter	Page

   Surveying, 1785
   Surveying, 1786
   Surveying, 1787-1788

VII. THE QUALITY AND COST OF THE FIRST SURVEYS 144

   The Quality of Surveying
   The Cost of Surveying

VIII. THE SIGNIFICANCE OF THE FIRST SURVEYS 155

   Influence upon Later Public Land Surveying
   Contribution to the Opening of the Northwest
   A Service to Mapping
   The Production of Historical Evidence

   PART III: INTERIM SURVEYING, AND FEDERAL
       SURVEYING RENEWED

IX. PRIVATE SURVEYING UNDER THE LAND ORDINANCE
   OF 1785 169

   The Last Days of Thomas Hutchins
   Survey of the Ohio Company Lands
   Survey of the Miami Purchase

X. HOW FEDERAL SURVEYING CAME TO BE RESUMED, UNDER THE
   LAND ACT OF 1796 185

   The Indians Sustain a Decisive Defeat
   Land Companies Lose Their Leadership
   Passage of the Land Act of 1796
   Federal Surveying Begins Again

XI. STATUS OF THE AMERICAN RECTANGULAR LAND SURVEY
   SYSTEM IN 1800 205

   Chain of Command
   The Contract System
   Surveying: Base Lines and Principal Meridians
   Surveying: General Procedure
   Survey Records

XII. SUMMARY OF FINDINGS 220

   The Rectilinear Grid
   Execution of Surveying
   The Founders
   Contributions to Mapping and Historical Knowledge
   Early Opinions for and against the Rectangular
   Survey System

APPENDIX 232

BIBLIOGRAPHY 235

INDEX 249

LIST OF ILLUSTRATIONS

Figure		Page
Frontispiece.	Extent of the American Rectangular Land Survey System	
1.	Claims in the Northwest by Virginia, Connecticut and Massachusetts	9
2.	Lands Reserved in Virginia's Deed of Cession	9
3.	Three Boundaries for Early Indian Cessions in the Northwest	12
4.	Projected Boundaries for Western States, 1784	18
5.	Projected Boundaries for Western States, 1784, Shown on Contemporary Map Base	27
6.	Lines Prescribed by the Northwest Ordinance for Bounding States	34
7.	Jeffersonian Units of Land Subdivision	47
8.	Solution to the Problem of Rectangles and Meridians	54
9.	Two Examples of Township-Bounding in New England	91
10.	Numbering of Townships and Lots under the Land Ordinance of 1785	99
11.	The First Scene of Survey and Its Environs	106
12.	The Seven Ranges	120
13.	Surveying Diagrams	147
14.	Ohio Company Lands and the Miami Purchase	171
15.	The Three Areas First Surveyed under Land Act of 1796	189
16.	Details of Surveying after Passage of Land Act of 1796	211

ERRATA

Page 3, fn. 5, line 7. For Malone Dumas read Dumas Malone.

Page 4, fn. 1. To follow blank after Jrnls. Cont. Cong.: XXVII, 453.

Page 9, Fig. 1. Across all of the Northwest should be written: Virginia's Claim.

Page 12, Fig. 2. Under Ft. Greenville, for Treaty, 1794 read Treaty, 1795.

Page 12, Fig. 2. In box, on Greenville Treaty Line, for 1794 read 1795.

Page 14, line 7. For Cayahoga read Cuyahoga.

Page 15, fn. 3, line 5. For Thompson read Thomson.

Page 16, line 26. Delete adult.

Page 18, Fig. 4b. Western boundary of North Carolina omitted.

Page 22, line 5. For not read now.

Page 32, line 5. Delete in mind.

Page 37, fn. 2, line 2. For Jnrls. read Jrnls.

Page 40, fn. 2, line 3. Quotes before title omitted.

Page 57, last paragraph, line 2. For charter read chapter.

Page 61, fn. 5. For Ohio Land Subdivisions read Original Ohio Land Subdivisions.

Page 70, fn. 2, line 1. For chap. iii read chap. iv.

Page 77, fn. 3, line 3. For Timothy read Abel.

Page 78, fn. 2, line 3. For Martinus read Martius.

Page 84, fn. 2, line 16. For Origins read Origin.

Page 106, Fig. 11. For Charlisle read Carlisle.

Page 108, line 3. For Harrisburgh read Harrisburg.

Page 113, line 11. For Federal read federal.

Page 114, fn. 5, line 6. For The State of Ohio read History of the State of Ohio.

Page 114, fn. 5, line 8. For 1942 read 1944.

Page 118, line 2 below the rule. For XI read XLI.

Page 122, fn. 1, line 6. For 1784 read 1884.

Page 125, lines 2-3. For Due to read Because of.

Page 136, line 17. For due to read because of.

viii

ERRATA--Continued

Page 139, line 12. For Absalom read Isaac.

Page 144, fn. 2, line 6. For 1876 read 1786.

Page 156, fn. 3, line 2. For secretaries read securities.

Page 157, fn. 2, line 4. For Tanges read Ranges.

Page 158, last footnote, line 1. For 5 read 4.

Page 158, last footnote, line 1. Add to the end of line: for.

Page 162, fn. 4, lines 5-6. For Biographical read Bibliographical.

Page 169, fn. 1, line 2. For chap. ix read chap. viii.

Page 173, fn. 1, line 1. To follow blank after p.: 21.

Page 179, fn. 2, line 3. For chap. ix read chap. viii.

Page 190, fn. 3. For Piankaskaw read Piankashaw.

Page 193, line 9. For nine read eleven.

Page 194, line 8. For Van Allen read Van Alen.

Page 196, fn. 4, line 5. To follow blank after p.: 74-75.

Page 203, line 17. For /superior/ 5 read /superior/ 6.

Page 209, fn. 4, line 7. Close parentheses at end of sentence.

Page 233, line 5. For 16 read 17.

Page 242, entry for Hosack, David. For Bibliographical read Biographical.

INTRODUCTION

As a striking example of geometry triumphant over physical geography, the general pattern of settlement in the vast area of the United States once owned by the national government (see Frontispiece) has often attracted the attention of writers on the American scene.[1] Underlying the prevailing tendency toward a geometrically regular disposition of roads, fields and street plans, in this area, are lines which were laid down by federal surveyors prior to the transfer of land from federal to private ownership. These lines, bounding "Congressional" townships, each six miles square, and component sections, each one mile square, represent the American rectangular land survey system.

The present study was prepared in order to increase our understanding of the American rectangular land survey system. Divided into three parts, the study consists of a report on the early years of the system's existence. The first part presents the original plan for national rectangular surveying--its background, authorship, content and altered embodiment in the Land Ordinance of 1785. The second part consists of a chronicle and analysis of public land surveying under the Ordinance, that is, of the survey of the "Seven Ranges." The third part opens with an account of continued rectangular surveying, by private land

---

[1] Wolfgang Langewiesche, for example, has called the rectilinear pattern of roads, oriented to the cardinal points of the compass, "one of the sights of the world," in his "The United States from the Air," Harper's Magazine, CCI (October, 1950), 188. See also, on the view from the air, Christopher Tunnard, "Fire on the Prairie," Landscape, II (Spring, 1952), 11, and John S. Radosta, "A New Yorker's Midwest Journal," New York Times, May 19, 1957, Sec. 2, Pt. II, p. 40. For the observations of two European geographers on settlement patterns in this area, see Hans Boesch, Amerikanische Landschaft, Neujahrsblatt der Naturforschenden Gesellschaft in Zurich auf das Jahr 1955 (Zurich, 1955), pp. 38-42, and Herbert Lehmann, "The Role of Law and Tradition in the Use of Agricultural Resources," Report of Seminar on Agricultural Utilization of Natural Resources, University of Chicago, Spring and Summer Quarters, 1952 (Mimeographed by Department of Geography, University of Chicago, November, 1952), pp. 2-3. Of the many examples of treatment of the same subject by American geographers, one of the most recent may be found in John H. Garland (ed.), The North American Midwest, A Regional Geography (New York: John Wiley and Sons, 1955), pp. 31-32.

companies, and proceeds through a discussion of the Land Act of 1796 and of the early years of federal surveying which followed passage of that Act, leading finally to a description of the status of rectangular surveying in 1800.

As the reader will discover, most of the pages in the study are taken up with a chronologically arranged narrative. This narrative, even where it broadens to include events indirectly associated with the main theme--development in the design and execution of surveying--comprises, together with accompanying descriptions, a simple factual record. To supplement the record and draw significance from it, interpretations are offered which explain, for example, the motives behind adoption of the idea of rectangular surveying, and the special contributions made by public land surveyors, incidental to their primary mission of preparing a wilderness for sale. For a summary of principal findings, the reader is referred to the concluding chapter of the study.

CHAPTER I

THREE PRIOR CONSIDERATIONS

In the beginning, March 1, 1784, Congress[1] accepted Virginia's cession of lands northwest of the Ohio River.[2] By this act, the national domain became a reality.[3] On the following day, the Secretary of Congress recorded the appointment of a committee "to devise and report the most eligible means of disposing of such part of the Western lands as may be obtained of the Indians by the proposed treaty of peace and for opening a land office."[4] Under the chairmanship of Thomas Jefferson, the following men served on the committee: Hugh Williamson of North Carolina, David Howell of Rhode Island, Elbridge Gerry of Massachusetts, and Jacob Read of South Carolina.[5] Gerry and Read later

---

[1] This was the Continental Congress, officially known, after adoption of the Articles of Confederation, as "The United States in Congress Assembled." The seat of Congress, from November 26, 1783, to June 3, 1784, was Annapolis, Maryland.

[2] Worthington C. Ford et al. (eds.), Journals of the Continental Congress, 1774-1789, Edited from the Original Records in the Library of Congress (34 vols.; Washington: Government Printing Office, 1904-1937), XXVI, 112-117. Hereafter referred to as Jrnls. Cont. Cong.

[3] "National domain" is used here in sense interchangeable with "public domain," to refer to lands in the ownership of the United States Government. For an exploration of the variety of meaning attaching to "public domain," see E. Louise Peffer, "Which Public Domain Do You Mean?" Agricultural History, XXIII (April, 1949), 140-146.

[4] Papers of the Continental Congress, CLXXVI, 151, General Records of the United States Government (Record Group 11), Foreign Affairs Section, Legislative, Judicial and Diplomatic Records Branch, National Archives, Washington. Hereafter referred to as Papers Cont. Cong. Initial number in every citation of Papers refers to a volume-group rather than to a single volume.

[5] Williamson, Howell and Read were serving their first (three year) terms in Congress. Gerry had served since 1776. For brief sketches of the careers of these men, see entries under their respective names in Biographical Directory of the American Congress, 1774-1949 (Washington: Government Printing Office, 1950). For additional biographical information, see Dictionary of American Biography, Allen Johnson and Malone Dumas (eds.) (20 vols.; New York: Charles Scribner's Sons, 1928-1937).

revealed their attitude toward the committee report by voting against its consideration by Congress.[1] Williamson, on the other hand, called the plan of survey in the report "our sheet anchor [which] is to be carefully managed."[2] Howell, by his own account, moved the appointment of the committee, and later gave evidence of support for its proposals.[3]

To turn immediately to the committee's proposed land ordinance would be to lose a convenient opportunity for presenting the background against which the committee worked. The preamble to the committee report stated three conditions which were to be satisfied before lands would be subject to survey and disposal: the lands were to be ceded by claimant states, purchased of the Indian inhabitants, and laid off into new states. Taking its cue from this preamble, the present chapter will provide a basis for further discussion by reviewing the first state cessions, describing the prospect for Indian cessions in the spring of 1784, and examining the plan for western states contained in the Ordinance of 1784, a companion law to the proposed ordinance on public lands.

State Cessions

The initial sequence of state cessions, a well-known story, may be dealt with briefly here.[4] To lands north and west of the Ohio River, Maryland, Delaware, New Jersey, Rhode Island

---

[1] Jrnls. Cont. Cong.

[2] Williamson to Governor of North Carolina, Edenton, North Carolina, July 5, 1784, in Edmund C. Burnett (ed.), Letters of Members of the Continental Congress (8 vols.; Washington: Carnegie Institution of Washington, 1921-1936), VII, 563. Hereafter referred to as Burnett, Letters of Members.

[3] Howell says he moved the appointment of the committee, in Howell to Jabez Bowen, Annapolis, March 12, 1784, in William R. Staples, Rhode Island in the Continental Congress (Providence, Rhode Island, 1870), p. 483. The Journals contain no mention of this motion. Howell's favorable attitude toward the proposed ordinance will be made apparent later in this study.

[4] State cessions north of the Ohio River are reviewed in B. A. Hinsdale, The Old Northwest (New York: Silver, Burdett and Company, 1888), pp. 188-246. On cessions both north and south of the Ohio River see Payson J. Treat, The National Land System, 1785-1820 (New York: E. B. Treat and Company, 1910), pp. 1-14, and Thomas Donaldson, The Public Domain, Its History, with Statistics, U. S. House Doc. No. 45, Pt. 4, 46th Cong., 3d Sess. (Washington: Government Printing Office, 1884), pp. 56-88.

ard New Hampshire made no claim. Pennsylvania's recognized rights embraced the headwaters of the Ohio River, but were confined by her present western boundary, a meridian five degrees of longitude west of a point on the Delaware River, in accordance with a settlement concluded with Virginia in 1779.[1] New York's present western boundary was accepted by Congress in 1782.[2] At the end of the Revolutionary War, the claims of three states were effective in the Northwest: those of Massachusetts, Connecticut and Virginia (Fig. 1). Massachusetts claimed westward to the Mississippi, between the northern and southern boundaries of a grant made to the Massachusetts Bay Company, in 1629.[3] Connecticut, on

---

[1] The agreement reached by commissioners of Pennsylvania and Virginia in 1779 and ratified by the two states in 1780 stated, "Mason and Dixon's line shall be extended due west five degrees of longitude, to be computed from the river Delaware, for the southern boundary of Pennsylvania forever." (Pennsylvania Archives, Third Series [30 vols.; Harrisburg, 1894-1899], III, 497.) The charter granted to William Penn in 1681 had set the western limit of Pennsylvania at five degrees west of the Delaware River. (Boyd Cumrine, "The Boundary Controversy between Pennsylvania and Virginia, 1748-1785," Annals of the Carnegie Museum, I [1901], 510.) Not until the agreement of 1779, however, was it officially affirmed that the boundary would take the form of a meridian, and that this meridian would be located five degrees west of that point on the Delaware where it was crossed by the Mason and Dixon Line.

[2] The New York law authorizing cession of claims west of the present state boundary was passed in 1780. For deed of cession, March 1, 1781, see Clarence E. Carter (ed.), The Territorial Papers of the United States, Vol. II: The Territory Northwest of the River Ohio, 1787-1803 (Washington: Government Printing Office, 1934), pp. 3-5. Hereafter referred to as Carter, Territorial Papers, II. For acceptance of deed by Congress, October 29, 1782, see Jrnls. Cont. Cong., XXIII, 694. New York's westerly claims were based not on charter rights but on an assumed sovereignty over the Indian "Five Nations." (Ibid., XXIII, 488.)

[3] Jrnls. Cont. Cong., XXIII, 488. The Mississippi River was accepted as a western limit by all three claimant states, in accordance with the treaty of peace with Great Britain. (Ibid., pp. 488-489.) Interpreting the Massachusetts charter of 1629, the legislature of Massachusetts declared, in 1784, that the southern limit of their claim was "a point or place situate in 42 degrees of northern latitude, 2 minutes north," and that the northern limit of their claim was "a place or point situate in 44 degrees northern latitude, 15 minutes north." (Journals of the American Congress, from 1774 to 1788 [4 vols.; Washington: Way and Gideon, 1823], IV, 445.) Parallels of latitude sweeping westward from these two points cut across the State of New York, and it was on account of a resulting dispute with New York that the latitude figures cited above were formally stated. Agents of New York and Massachusetts made a settlement of conflicting claims, December 16, 1786. (Ibid., pp. 787-790.)

the basis of a grant to the Connecticut Company, in 1662, claimed a parallel strip, bounded by the Massachusetts claim on the north and the forty-first parallel of latitude on the south.[1] Virginia's claim, in contrast, was not confined by parallels of latitude, but swept northwestward throughout the territory beyond the Ohio River, on the authority of a grant of 1609.[2] During the Revolutionary War, expeditions of Virginians into the Northwest, undertaken for the prevention of Indian attacks on the Kentucky settlements, had been crowned by the success of George Rogers Clark, in the capture of Vincennes and French villages in the Illinois country.[3]

---

[1] Jrnls. Cont. Cong., XXIII, 488. The charter of 1662 bounded Connecticut on the north by the Massachusetts line, and on the south by the sea. (Journals of American Congress, IV, 135.) In 1683, an agreement was reached with New York, which anchored the western boundary of Connecticut along the coast of Long Island Sound, at the mouth of the Biram River. (Ibid., p. 136.) This river-mouth, at almost exactly forty-one degrees north latitude, in turn became the reference point for the southern limit of Connecticut's claims. Between the forty-first parallel on the south and the latitude of the Massachusetts boundary on the north, Connecticut claimed westward, skipping over the Hudson River Valley because of the agreement with New York. During the quarter of a century which preceded the Revolutionary War, Connecticut settlers, moving westward within these bounds, settled in the Lehigh Valley of Pennsylvania and organized the county of Westmoreland. A bitter controversy between Connecticut and Pennsylvania ensued, which was settled in favor of Pennsylvania by a court appointed under the Articles of the Confederation, in 1782. (Ibid., p. 140.) Deprived of her first western claim, Connecticut was intent upon pressing her claim to lands beyond Pennsylvania. See William Samuel Johnson to Roger Sherman, New York, April 20, 1785, in Burnett, Letters of Members, VIII, 102.

[2] Jrnls. Cont. Cong., XXIII, 489. Virginia's charter of 1609 set aside all the territory two hundred miles northward and two hundred miles southward of Old Point Comfort, "and all that Space and Circuit of Land, lying from the Sea Coast of the Precinct aforesaid, up into the Land throughout from Sea to Sea, West and Northwest." Francis N. Thorpe, The Federal and State Constitutions, Colonial Charters, and Other Organic Laws of the States, Territories and Colonies Now or Heretofore Forming the United States of America, U.S. House Doc. No. 357, 59th Cong., 2d Sess. [7 vols.; Washington: Government Printing Office, 1909], VII, 3795.)

[3] On this campaign, by which Clark has often been said to have won the Northwest for the United States, see James A. James, The Life of George Rogers Clark (Chicago: University of Chicago Press, 1928), pp. 109-146. Clark and his men first captured the Illinois villages of Kaskaskia, Prairie du Rocher, St. Phillipe and Cahokia, July 4-6, 1778. (Ibid., pp. 119-121.) Vincennes was captured from the British February 24, 1779. (Ibid., pp. 142-145.) "This stroke," said Clark, "will nearly put an end to the Indian War, had I but men enough to take the advantage of the

Virginia's extensive western claims had comprised the particular grounds for Maryland's celebrated refusal to ratify the Articles of the Confederation, in 1777, without Congress' being given the power to "ascertain and restrict the boundaries of such of the confederated states which claim to extend to the River Mississippi or the South Sea."[1] In 1779, Virginia passed a land act whose surveying content will be a subject of repeated reference in forthcoming pages.[2] This act threw open Kentucky lands for sale, but notably excluded lands north of the Ohio, foreshadowing their cession to the United States in response to the protests of Maryland and other landless states. Subsequently, in January, 1781, Virginia offered to cede the Northwest, and Maryland soon after entered the Union.[3] For three years, acceptance

---

present confusion of the Indian Nations, I could silence the whole in two months." (Clark to Benjamin Harrison, Vincennes, March 10, 1779, in James A. James [ed.], George Rogers Clark Papers, 1771-1781, Collections of the Illinois Historical Library [Springfield, Illinois: Illinois State Historical Library], VIII [1912], 305.) An attack on Detroit, which Clark had in mind, never materialized.

[1] Herbert B. Adams, Maryland's Influence Upon Land Cessions to the United States, Johns Hopkins University Studies in Historical and Political Science, Third Series, No. 1 (Baltimore: Johns Hopkins University, 1885), p. 22. The thesis of this study is that Maryland should be credited with bringing about national ownership of western lands.

[2] Two general land laws were enacted by Virginia in 1779, one for settling titles to lands already claimed, and the other for opening a land office. The latter law, referred to here, may be found in William W. Hening (ed.), The Statutes at Large; Being a Collection of All the Laws of Virginia, from the First Session of the Legislature, in the Year 1619 (13 vols.; Richmond, Virginia, 1819-1823), X, 50-65. Hereafter referred to as Hening, Virginia Statutes.

[3] Events leading up to Virginia's first offer of cession, January 2, 1781, are analyzed in Merrill Jensen, "The Cession of the Old Northwest," Mississippi Valley Historical Review, XXIII (June, 1936), 27-48. Jensen stresses the influence of land companies and calls into question Adams' view that Maryland was motivated by a vision of national welfare in insisting upon the cession of western claims. Maryland's ratification of the Articles of Confederation is attributed to the desire of France for completion of the Confederation in St. George L. Sioussat, "The Chevalier de La Luzerne and the Ratification of the Articles of the Confederation," Pennsylvania Magazine of History and Biography, LX (October, 1936), 391-418. The complexities of land interests and politics involved in Virginia's first act of cession are further explored in Thomas P. Abernathy, Western Lands and the American Revolution (New York: D. Appleton-Century Company, 1937), pp. 116-247.

of Virginia's cession remained in abeyance, a period marked by conflict and intrigue arising from the attempts of private land companies to gain recognition of their title to land within Virginia's jurisdiction.[1] The period of waiting was brought to a close when Congress signified substantial acceptance of Virginia's conditions.[2] Virginia made a second offer of cession in December, 1783, Congressional acceptance of which was accomplished March 1, 1784.[3]

The national domain soon filled out the boundaries of the Northwest, with the completion of a cession by Massachusetts in 1785,[4] and another by Connecticut in the following year.[5] Islands of reserved land remained, however. Connecticut's cession was employed as a means of confirming her title to a remnant of her claim, a "western reserve" lying between the forty-first parallel and Lake Erie, and extending one hundred and twenty miles west from the Pennsylvania line (Fig. 1).[6] Though Connecticut later surrendered political jurisdiction, the soil of the reserve remained outside the public domain.[7] The Virginia deed of cession created several islands of reserved land. First, it required that the national government confirm the titles of "the French and

---

[1] This period of waiting is interpreted by Merrill Jensen, in his "The Cession of the National Domain, 1781-1784," *Mississippi Valley Historical Review*, XXVI (December, 1939), 323-342.

[2] *Ibid.*, pp. 336-337.

[3] *Ibid.*, p. 341. For the committee report which cleared the way for Virginia's second cession, see *Jrnls. Cont. Cong.*, XXV, 559-563.

[4] The Massachusetts deed of cession, April 19, 1785, appears in Carter, *Territorial Papers*, II, 10-12. See also *Jrnls. Cont. Cong.*, XXVIII, 281-283. For acceptance by Congress, *ibid.*, p. 283.

[5] The Connecticut deed of cession, September 13, 1786, appears in Carter, *Territorial Papers*, II, 22-24. See also *Jrnls. Cont. Cong.*, XXXI, 654-655. For acceptance by Congress, *ibid.*, p. 655.

[6] Carter, *Territorial Papers*, II, 14; *Jrnls. Cont. Cong.*, XXXI, 654.

[7] Connecticut's offer of cession of jurisdiction, December 31, 1798, appears in Carter, *Territorial Papers*, II, 657-658. Connecticut had sold her title to land in the reserve to a private land company in 1785. For act of Congress completing transfer of jurisdiction, see Carter, *Territorial Papers*, Vol. III: *The Territory Northwest of the River Ohio, 1787-1803, Continued* (Washington: Government Printing Office, 1934), p. 84.

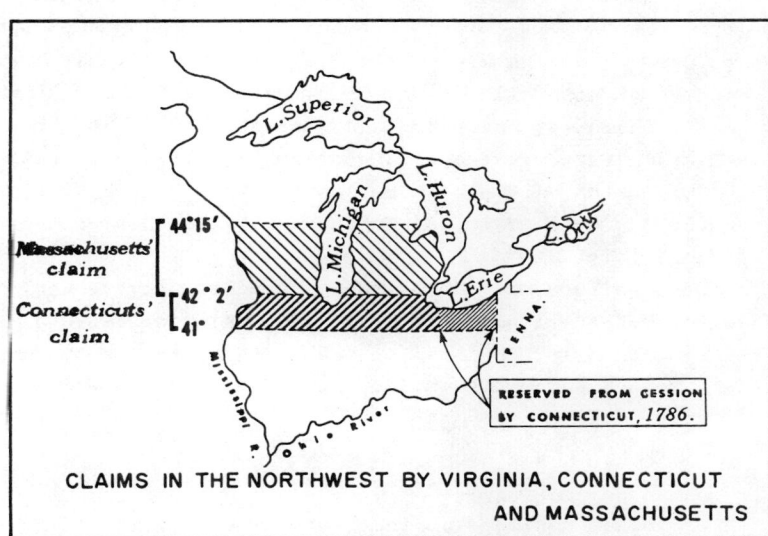

Fig. 1

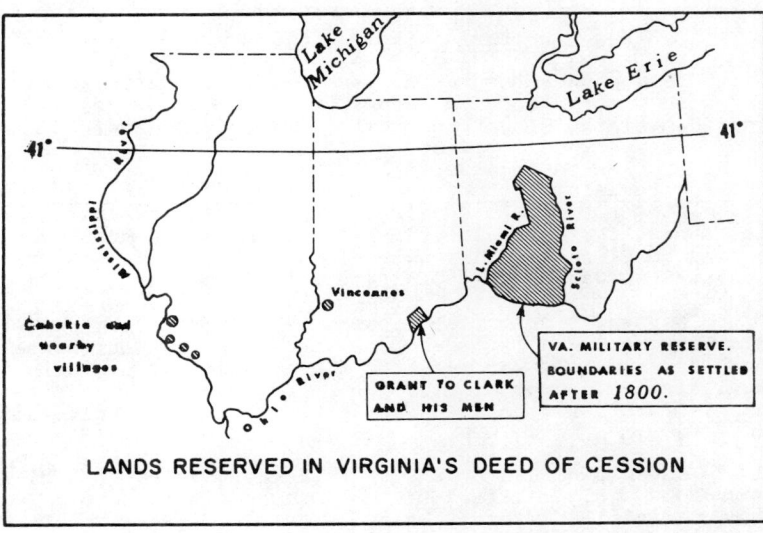

Fig. 2

Canadian and other Settlers" at Vincennes and in the Illinois country, who had professed themselves citizens of Virginia (Fig. 2).[1] Second, it allowed one hundred and fifty thousand acres to General George Rogers Clark and his men.[2] This award was laid off at the falls of the Ohio, opposite Louisville (Fig. 2).[3] Third, it reserved lands between the Scioto and the Little Miami rivers, in the southwestern part of the present state of Ohio, for Virginia's Continental troops, should bounty lands reserved for them in the Kentucky Military District prove insufficient (Fig. 2).[4]

When Jefferson's committee convened to provide for the disposal of western lands, the national domain consisted of Virginia's cession only, confined on the north by the forty-first parallel, beyond which Connecticut and Massachusetts still maintained their claims.

Indian Cessions

Of Indian cessions in the West, at this time, there were none. In the treaty of peace with Great Britain, ratified by Congress in January, 1784,[5] the Indians were not mentioned, nor had they been consulted about the treaty.[6] Great Britain had

---

[1] Carter, Territorial Papers, II, 8.   [2] Ibid.

[3] This area, conspicuous because its subdivision does not accord with the surrounding federal rectangular surveys, lies in present-day Clark, Floyd and Scott counties, Indiana. Clark had already received, as a private individual, a grant of this land from the Piankeshaw Indians in 1779. (American State Papers: Documents, Legislative and Executive, of the Congress of the United States [38 vols.; Washington: Gales and Seaton, 1832-1861], Public Lands, I, 247.) The Virginia assembly had granted the same land to Clark's regiment, and a board to survey the tract was created in October, 1783. (Abernathy, op. cit., p. 296.)

[4] Carter, Territorial Papers, II, 8. The deed of cession should have reserved land for both the State and Continental troops of Virginia. For notes on the omission of State troops through clerical error, see Julian P. Boyd (ed.), The Papers of Thomas Jefferson (Princeton, New Jersey: Princeton University Press, 1950-   ), IV, 390, n. and VI, 580, n.

[5] The Definitive Treaty of Peace was signed at Paris, September 3, 1783, ratified by the United States January 14, 1784, and ratified by Great Britain April 9, 1784. The text of the treaty appears in Hunter Miller (ed.), Treaties and Other International Acts of the United States of America (Washington: Government Printing Office, 1931-   ), II, 151-157.

[6] The discontent which prevailed among the Indians when they learned of the treaty is described in a letter from Brigadier

yielded sovereignty to the United States over lands extending from the thirty-first parallel of latitude to the Great Lakes, and westward to the Mississippi River, but the terms of peace with western Indian tribes had yet to be negotiated. The committee on public lands, it will be recalled, was expected to devise a way of disposing of "such part of the Western lands as may be obtained of the Indians by the proposed treaty of peace."[1]

The Congressional view of the proposed treaty with the Indians, as of the beginning of March, 1784, may be judged from the content of the report of a committee on Indian affairs, made in the previous October.[2] This report anticipated "establishing a boundary line of property for separating and dividing the settlements of the citizens from the Indian villages and hunting grounds, . . . thereby extinguishing as far as possible all occasion for future animosities, disquiet and contention."[3] The boundary aimed for would have been placed far enough west to include Virginia's military reserve. It would have followed up the Great Miami River from its mouth to its confluence with the Mad River, thence northwest to the portage at the head of the Maumee River, and down that river to Lake Erie, thus setting off nearly all of the present state of Ohio (Fig. 3).[4] This was to be a dictated settlement, in consideration of the obligation the Indians were under "to make atonement for the enormities they have perpetrated, and a reasonable compensation for the expenses which the United States have incurred by their wanton barbarity."[5] The

---

Maclean, a British officer at Niagara, to General Frederick Haldimand, May 18, 1783, in "Haldimand Papers," <u>Michigan Pioneer and Historical Society Collections</u>, XX (1892), 117-121.

[1] Papers Cont. Cong., CLXXXVI, 151.

[2] Report of committee on Indian affairs, October 15, 1783, in <u>Jrnls. Cont. Cong.</u>, XXV, 681-695.

[3] <u>Ibid.</u>, p. 686.

[4] The boundary was described as follows: "Beginning at the mouth of the Great Miami River, which empties into the Ohio, thence along the said river Miami to its confluence with the Mad river; thence by a direct line to the Miami fort at the village of that name on the other Miami river, i.e., the Maumee River which empties into Lake Erie, comprehending all the lands between the above mentioned lines and the State of Pensylvania on the East, Lake Erie on the North and the River Ohio on the South East." <u>Jrnls. Cont. Cong.</u>, XXV, 686. The "Miami fort at the village of that name" was approximately at the site of Fort Wayne, as shown, Fig. 3.

[5] <u>Ibid.</u>, p. 683.

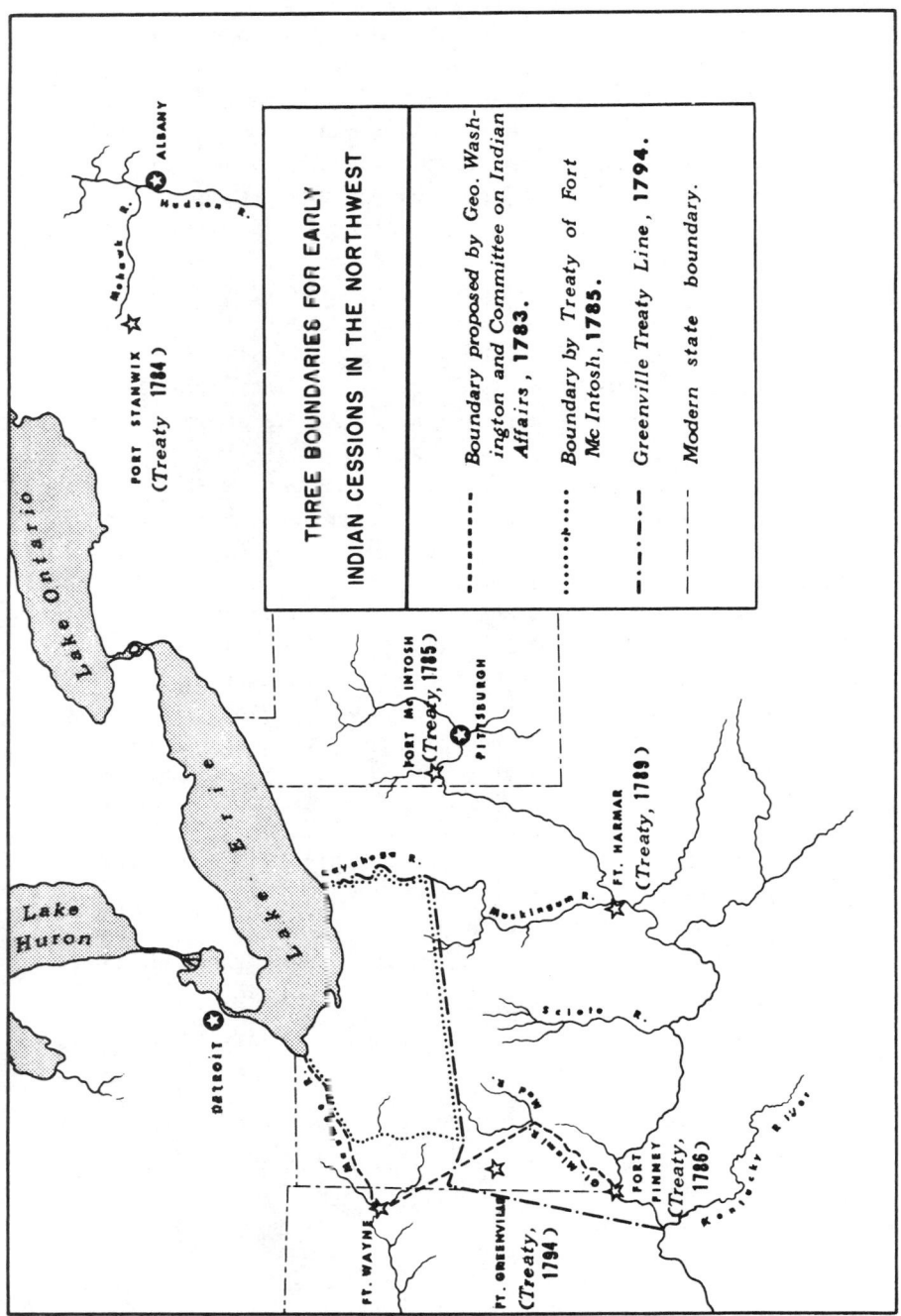

Fig. 3

lands to be ceded by the Indians were to be purchased only in the sense that presents of "coarse goods, . . . cloathing, and other articles" would be given.[1] Commissioners were expected to gain this territory through a single treaty with "the Indians inhabiting the . . . [lands] and their allies and dependents."[2]

Meanwhile, the Indians of the Northwest had taken measures to form a confederacy. In September, 1783, a grand council was held at Sandusky, under the leadership of the Six Nations (Iroquois) of western New York state, with the encouragement of the British, who had retained possession of forts at Detroit and Niagara. Here it was solemnly agreed that the Ohio River should be the boundary between the red man and the white forever.[3] Perhaps with knowledge of this threat of unified resistance, a new Congressional committee on Indian affairs, reporting March 19, 1784, abandoned the idea of a single treaty, and adopted a policy of divide et impera. They instructed treaty commissioners, newly appointed, to treat with the tribes "at different times and places" and to "discourage every coalition."[4]

Jefferson, Williamson and Howell, already introduced here as members of the committee on public lands, made up this new committee on Indian affairs. Their additional proposal that the Indians be pushed back to "a meridian line passing through the lowest point of the rapids of [the] Ohio"[5] never found a counterpart in reality, but the principle of separate treaties, which they enunciated, became the basis of later federal policy. After many delays, United States commissioners succeeded in dictating terms to the Six Nations, by the Treaty of Fort Stanwix, dated October 22, 1784.[6] The Six Nations, in this treaty, gave up all

---

[1] Ibid., p. 688. Congress voted $15,000 for this purpose, March 19, 1784. (Ibid., XVI, 154.)

[2] Ibid., XV, 687.

[3] The proceedings of the council at Sandusky, August 26-September 8, 1783, are printed in "Haldimand Papers," Michigan Pioneer and Historical Collections, XX (1892), 174-183.

[4] Jrnls. Cont. Cong., XXVI, 152-154.

[5] Ibid., p. 153.

[6] Preparations for the treaty and complications arising from New York's separate negotiations with the Indians are described in Henry S. Manley, The Treaty of Fort Stanwix, 1784 (Rome, New York: Rome Sentinel Company, 1932), pp. 14-89. Rome, New York, in the upper Mohawk Valley, now covers the site of Fort Stanwix.

claim to lands west of Pennsylvania.[1] Shifting the scene of negotiation from the Mohawk Valley of New York to the upper Ohio River, just within the Pennsylvania line,[2] American commissioners[3] next met with representatives of four tribes of the Ohio country, the Wyandot, Delaware, Chippewa and Ottawa. By the Treaty of Fort McIntosh, January 21, 1785, the Wyandot and Delaware were allowed a broad zone along Lake Erie, from the Cayahoga to the Maumee (Fig. 3), but they were obliged to yield all other claim to the Ohio country, and even within their reserve strategic points were yielded.[4]

Seemingly, the opening of the Northwest to survey and settlement was well begun. But a British observer at Detroit foresaw that "the transactions at those two Meetings can not be permanent, as it will be found that refractory tribes will never tamely submit to be deprived of A Country on which they think their existence depends."[5] This anticipation was fulfilled in December, 1786, when the Indian confederacy met at Detroit and renounced the treaties, declaring, "You kindled your council fires when you thought proper, without consulting us, . . . and have entirely neglected our plan of having a general conference with the different nations of the Confederacy."[6] Three additional separate

---

[1] The text of the treaty appears in *American State Papers*, *Indian Affairs*, I, 10. See also *Jrnls. Cont. Cong.*, XXVIII, 423-424.

[2] Fort McIntosh, the scene of the second treaty, was located on a bluff on the north side of the Ohio River, below the mouth of Big Beaver Creek (Fig. 3).

[3] Richard Henry Lee, delegate to Congress from Virginia, and General Richard Butler served as commissioners at both Fort Stanwix and Fort McIntosh. General George Rogers Clark, unable to serve at Fort Stanwix, where General Oliver Wolcott took his place, was the third commissioner at Fort McIntosh.

[4] The reserve was allotted to "the Wyandot and Delaware nations, to live and hunt on, and to such of the Ottawa nation as now live thereon." The strategic points, set aside for trading posts, were at Sandusky, the lower rapids of the Sandusky River, the portage of the Great Miami (i.e., the southwest corner of the reserve), and the mouth of the Maumee. See text of treaty in *American State Papers*, *Indian Affairs*, I, 11. See also *Jrnls. Cont. Cong.*, XXVIII, 424-426.

[5] Alexander McKee to Sir John Johnson, April 24, 1785, quoted in Randolph C. Downes, *Council Fires on the Upper Ohio: A Narrative of Indian Affairs in the Upper Ohio Valley until 1795* (Pittsburgh: University of Pittsburgh Press, 1940), p. 293.

[6] *American State Papers*, *Indian Affairs*, I, 9.

treaties failed to halt Indian incursions into ceded lands, which led to open war, 1790-1794.[1] A general settlement, including an effective cession line (Fig. 3), was not reached until the signing of the Treaty of Greenville, in 1795, following General Wayne's victory at Fallen Timbers.[2]

In March, 1784, Congress could not know the difficulties which would arise in making the first Indian cessions secure. Judging from correspondence, there would seem to have been greater confidence in the future of Indian cessions in the West than in that of cessions from the remaining claimant states.[3]

Jefferson's Plan for Western States

On the day that Virginia's deed of cession was accepted, Jefferson, heading a committee of three, laid before Congress an ordinance for government in the western territories.[4] Taking as its area of reference not only the Northwest but all of the western lands, ceded and unceded, the committee report as amended and finally approved by Congress, April 23, 1784, came to be known simply as the Ordinance of 1784.[5] It has been called the yoke-

---

[1] Treaties were concluded with the Shawnee at Fort Finney (at the mouth of the Great Miami River), January 31, 1786; with the Chippewa and other tribes at Fort Harmar (at the mouth of the Muskingum River), January 9, 1789; and with the "Six Nations" at Fort Harmar, January 9, 1789. For an account of American aggression, and the resulting war waged by Indian tribes to restore the Ohio River as a boundary, see Downes, op. cit., pp. 294-338.

[2] For text of the Treaty of Greenville, signed August 3, 1795, see American State Papers, Indian Affairs, I, 562-563. A Wyandot chief, at the treaty conference, said that heretofore the Fifteen Fires had been addressed as "Brother," but that he and other chiefs present "do now, and will henceforth, acknowledge the fifteen United States of America to be our father . . . [we] must call them brothers no more." Ibid., p. 580.

[3] Compare, for example, the matter-of-fact attitude toward Indian cessions on the part of Gerry, Jefferson and Sherman, in Burnett, Letters of Members, VII, 461, 462 and 479, respectively, with the uncertainty respecting further state cessions expressed by Spaight, Thompson and Williamson, ibid., pp. 510, 531, and 596, respectively.

[4] The other two members of the committee were Jeremiah T. Chase, of Maryland, and David Howell, of Rhode Island. Jrnls. Cont. Cong., XXVI, 118. For text of the report, ibid., pp. 118-120.

[5] For text of the final Ordinance, see ibid., pp. 275-278. The Ordinance may also be found in Boyd, Papers of Thomas Jefferson, VI, 613-615. Boyd thoroughly examines the background of the

mate of the proposed land ordinance of 1784.[1]

Of the Ordinance of 1784 a recent historian has written:

> It is too often said, and believed, that the Northwest Ordinance of 1787, which repealed the Ordinance of 1784, provided for democracy in the territories of the United States. The reverse is actually true. Jefferson's Ordinance provided for democratic self-government of western territories and for that reason it was abolished in 1787 by . . . [those who] wanted congressional control of the West so that their interests could be protected from the actions of the inhabitants.[2]

The critical difference between the two laws, to which this passage refers, concerned the proposed terms of government prior to statehood. Under the Ordinance of 1784, free adult males of a district scheduled to become a state were authorized to meet and form a temporary government, adopting "the constitution and laws of any one of the original states," and proceeding to the election of their own legislature. This government was to remain in effect until "such state shall have acquired twenty thousand free inhabitants," when a convention was to be called, establishing a permanent government and constitution.[3] Under the Ordinance of 1787, temporary government was to be conducted by a governor, a secretary, and three judges appointed by Congress. Even after the election of an assembly, in a second stage of government, the governor was to have an absolute veto on all legislation. Eligibility for statehood was to be delayed until the population rose to sixty thousand adult inhabitants.[4] The central concept of both laws was that new states, republican in form, should be admitted in due course into the Union, on a basis of equality with the original states.

First in the order of procedure outlined in the Ordinance of 1784 was the demarcation of state boundaries. These were to

---

Ordinance, ibid., pp. 581-600, and reproduces, with notes, four documents in the line of development of the Ordinance, ibid., pp. 600-613.

[1] A brief account of the concurrent consideration of these two ordinances in Congress appears in Edmund C. Burnett, The Continental Congress (New York: The Macmillan Company, 1941), pp. 598-601.

[2] Merrill Jensen, The New Nation: A History of the United States During the Confederation, 1781-1789 (New York: Alfred A. Knopf, 1950), p. 354.

[3] Jrnls. Cont. Cong., XXVI, 276.

[4] Ibid., XXXII, 335-342; also, Carter, Territorial Papers, II, 41-49.

be formed as follows:

> . . . by parallels of latitude, so that each State shall comprehend from north to south two degrees of latitude, beginning to count from the completion of forty-five degrees north of the equator; and by meridians of longitude, one of which shall pass through the lowest point of the rapids of Ohio, and the other through the western cape of the mouth of the Great Kanhaway [Kanawha]: but the territory eastward of this last meridian, between the Ohio, Lake Erie and Pennsylvania, shall be one State whatsoever may be its comprehension of latitude. That which may lie beyond the completion of the 45th degree between the said meridians, shall make part of the State adjoining it on the south: and that part of the Ohio, which is between the same meridians coinciding nearly with the parallel of $39°$ shall be substituted so far in lieu of that parallel as a boundary line.[1]

Specifications in this final form (Fig. 4B) differed from those in Jefferson's initial report (Fig. 4A) in omitting state names, accepting the Ohio River as a boundary for the part of its course between the two prescribed meridians, and reckoning from the forty-fifth parallel southward rather than from the thirty-first parallel northward. The names could be easily parted with. The Ohio River had already won acceptance as a boundary in the part concerned.[2] The reversal of the order of reckoning suggests that the northern tiers of states, which alone had been given names and were now deprived of them, were meant by this means to be restored to priority of attention. These were, of course, the states of the Virginia cession.

Since a modern, accurate map-base (Figs. 4A and 4B) necessarily falsifies the picture of the West upon which Jefferson, and those who concurred in his proposal, projected the stipulated parallels and meridians, an additional illustration, based upon a map by Thomas Hutchins, dated 1778, has been prepared (Fig. 5).[3]

---

[1] Jrnls. Cont. Cong., XXVI, 275-276. It should be added that these boundaries were not dictated unreservedly, but were expected to take state and Indian cessions into account, conforming to this scheme as nearly as such cessions would admit. Ibid., p. 275.

[2] Boyd points out that, just as Jefferson's thirty-ninth parallel was abandoned in favor of the Ohio River, which was the recognized northern boundary of the Kentucky country, so "Pelisipia" was the first of the proposed state names to be discarded by Congress, it being redundant since "Kentucke" was already an accepted name for this area. (Boyd, Papers of Thomas Jefferson, VI, 596-598.) Boyd suggests that this initial rejection, on the grounds of conflict with established usage, led to removal of all the names, and that dislike of the names in themselves was not, as has been supposed, the determining reason for their rejection.

[3] Thomas Hutchins, A New Map of the Western Parts of

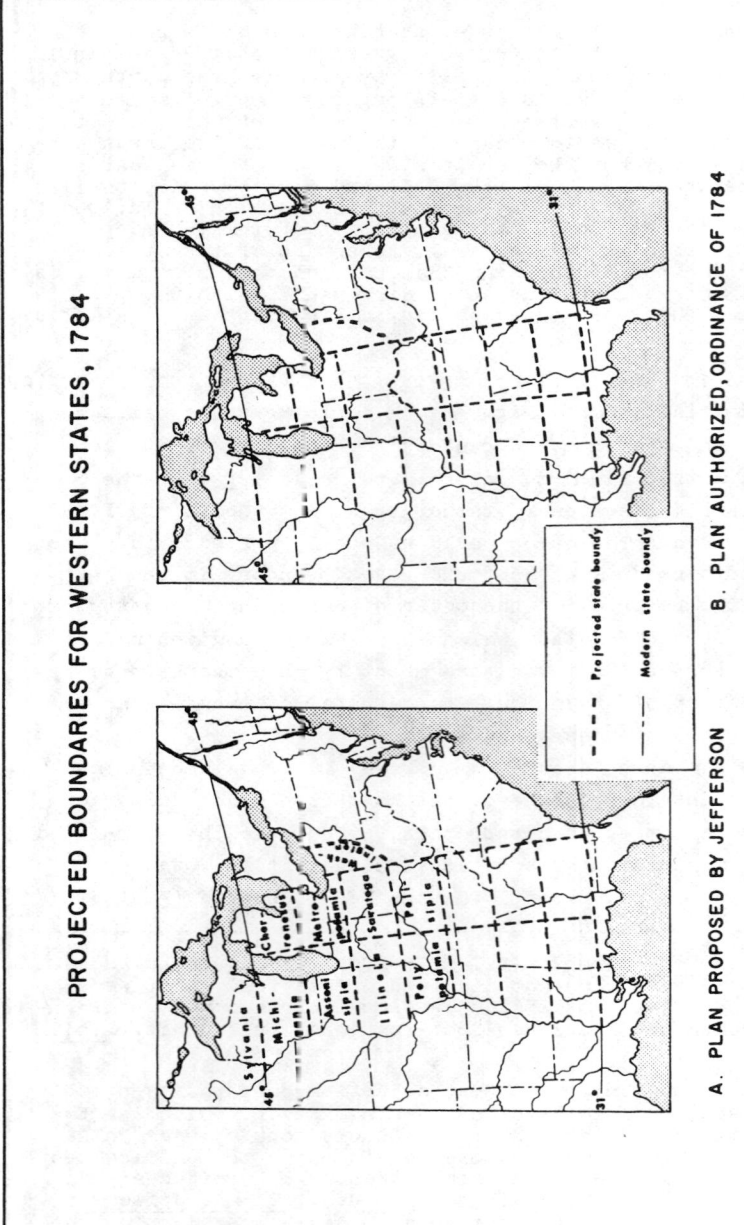

Fig. 4

That Jefferson accepted Hutchins' map as authoritative is strongly suggested by the fact that he borrowed its contents "on the western side of the Alleganey" for a map which he compiled in 1786, to accompany his Notes on Virginia.[1]

In the magnitudes of the proposed states, as in other respects, the Ordinance of 1784 was consistent with previously expressed Congressional intentions. By a resolution of October 10, 1780, new states were to measure "not less than one hundred nor more than one hundred and fifty miles square, or as near thereto as circumstances will admit."[2] Later reaffirmed, this condition was included in Virginia's deed of cession.[3] Even closer to the mark was a passage in a resolution offered before Congress in

---

Virginia, Pennsylvania, Maryland and North Carolina; Comprehending the River Ohio, and All the Rivers, Which Fall into It; Part of the River Mississippi, the Whole of the Illinois River, Lake Erie; Part of Lakes Huron, Michigan &c. And all the Country Bordering on these Lakes and Rivers. London: engraved by I. Cheevers, 1778. Scale: 1 inch to about 20 miles. To accompany the map, a pamphlet was prepared: Thomas Hutchins, A Topographical Description of Virginia, Pennsylvania, Maryland, and North Carolina, Comprehending the Rivers Ohio, Kenhawa, Sioto, Cherokee, Wabash, Illinois, Mississippi, &c . . . (London, 1778). Jefferson's copy of the pamphlet is in the Rare Books Division, Library of Congress. Jefferson is known to have purchased a copy of the map, as well. (Boyd, Papers of Thomas Jefferson, V, 585, n.) In lieu of Jefferson's copy of the map, not found, Fig. 5 has been based upon another copy, in the Map Division, Library of Congress.

[1] "A Map of the country between Albemarle Sound, and Lake Erie, comprehending the whole of Virginia, Maryland, Delaware, and Pennsylvania, with parts of several other of the United States of America," in Paul Leicester Ford (ed.), The Writings of Thomas Jefferson (10 vols.; New York: G. P. Putnam's Sons, 1892-1899), III, foll. p. 84. Scale: 1 inch to about 20 miles. Acknowledgment of the use of Hutchins' map appears on the face of this map. See also Jefferson to Col. William Stephen Smith, Paris, August 9, 1786, in Boyd, Papers of Thomas Jefferson, X, 213. Jefferson exchanged letters with Hutchins on the subject of Hutchins' map shortly before the Ordinance of 1784 was drawn up. See Hutchins to Jefferson, Philadelphia, February 11, 1784, ibid., VI, 535-536. It is worth noting that the expression "Rapids of Ohio," which appears on Hutchins' map in place of the usual "Falls of the Ohio," also appears in the ordinance as reported by Jefferson's committee. (Ibid., VI, 603.)

[2] Jrnls. Cont. Cong., XVIII, 915. Jefferson's "Saratoga" (Fig. 4), measured about one hundred forty miles by two hundred miles.

[3] For this condition in resolution of September 13, 1783, see Jrnls. Cont. Cong., XXV, 560, and for the same in Virginia's deed of cession, Carter, Territorial Papers, II, 7.

1783.[1] This proposed law, intended to accompany an acceptance of Virginia's first act of cession, contained the "Financier's Plan" for western settlement, whereby the expenses of government would have been paid out of income from reserved lands. Incidental to the plan was a provision that "the said territory shall be laid off in districts not exceeding two degrees of latitude and three degrees of longitude each," these districts to become states when sufficiently populated.[2]

Parallels of latitude already divided the West, as a glance at any map of state claims will confirm. Jefferson's state boundaries coincided, perhaps by design, with these prior lines in two instances: along the thirty-fifth parallel (the southern boundary of present-day Tennessee), which separated the respective claims of North and South Carolina, and along the forty-first parallel, which limited Connecticut's claim on the south. The "ladder" of Jefferson's odd-numbered parallels was completed on the south by the thirty-first parallel, which comprised the southern boundary of the United States, by the recent treaty of peace with Great Britain.[3] The survey of such lines posed two problems, that of determining the latitude of a beginning point, and that of prolonging a line while maintaining a constant latitude. Latitudes of occupied points could be determined with reasonable accuracy by qualified Americans, at this time. For example, Hutchins located the beginning point of the public land surveys, in 1785, with an error of less than one-half minute.[4] That a parallel of latitude could be prolonged satisfactorily was demonstrated by a group of Pennsylvania and Virginia commissioners, who, in the summer of 1784, extended Mason and

---

[1] Resolution moved by Theodoric Bland, of Virginia, June 5, 1783, in Jrnls. Cont. Cong., XXIV, 384-386.

[2] Ibid., p. 385. Jefferson's "Saratoga" measured two degrees of latitude by about three and one-half degrees of longitude.

[3] Hunter Miller, op. cit., II, 153. This parallel of latitude, which comprised the boundary from the Mississippi River eastward to the Chattahoochee River, was not accepted by Spain as a northern boundary of her province of West Florida until 1795. For agreement with Spain, October 27, 1795, see American State Papers, Foreign Relations, I, 547.

[4] Hutchins reckoned the latitude of this point to be 40°38'02" North. (Hutchins to the President of Congress, New York, November 24, 1785, Papers Cont. Cong., LX, 194.) The modern calculation is 40°38'27" North.

Dixon's line westward to the southwestern corner of Pennsylvania.[1]

For the use of meridians as state boundaries there was ample precedent. A meridian bounded Maryland on the west under its original charter.[2] New York had ceded all claim west of "a meridian Line to be drawn . . . through the most westerly Bent or Inclination of Lake Ontario,"[3] and Pennsylvania had accepted a meridian for a boundary, whose partial running, in the summer of 1785, would provide a point of departure for the public land surveys.[4] Jefferson's two meridians promised to be relatively easy to survey, since the position of each, like that of the meridian bounding New York, was defined by a landmark.[5] The extension of a meridian through a given point is perhaps the simplest kind of assignment for the running of a lengthy terrestrial line. The reputation of meridians for difficulty of survey derives from a separate problem, that of establishing longitude relative to a distant point of reference. Requiring meridians to be run seven degrees and ten degrees west of Philadelphia, for example, would have meant the necessity of establishing an accurate chronological

---

[1] The commissioners extended Pennsylvania's southern boundary from the western end of Mason and Dixon's line for a distance of about twenty-two miles to the southwest corner of the state. The eastern and western ends of this extension were found, in 1883, to differ in latitude less than one second, that is, less than one hundred feet. ("Report on the Survey of the West Virginia and Pennsylvania Boundary, 1883," Report of the Secretary of Internal Affairs of the Commonwealth of Pennsylvania, Containing Reports of the Surveys and Re-Surveys of the Boundary Lines of the Commonwealth [Harrisburg, 1887], pp. 396-402.) See also Map No. 4 and Map No. 5, ibid., Map Supplement.

[2] The charter to Lord Baltimore, 1632, set the western border at "the true meridian of the first fountain of the River Pattowmack." (F. N. Thorpe, op. cit., III, 1678.) Though the exact position of this meridian remained long in dispute, Virginia accepted this boundary in principle in its constitution of 1776. (Ibid., VII, 3818.)

[3] Carter, Territorial Papers, II, 11.

[4] The running of Pennsylvania's western boundary is discussed in chap. iv, below.

[5] New York's deed of cession offered two alternative landmarks as points of reference for a western state boundary. The "most westerly Bent or Inclination of Lake Ontario" was to determine the bounding meridian only if such a meridian proved to be "Twenty Miles due West from the most westerly Bent or Inclination of the River or Strait of Niagara." If the meridian was found to fall short of that distance, then the point designated on the Niagara River was to become the determining landmark, twenty miles west of which a meridian would be surveyed. (Carter, Territorial Papers, II, 4-5.)

relationship between widely separated points.[1] Requiring meridians simply to be located by such points as the edge of a river mouth and the lower end of a rapids removed this difficulty altogether. The wider significance of Jefferson's two meridians will not be considered in detail.

The first of the two meridians proposed, that passing through the western edge of the mouth of the Kanawha River (Figs. 4, 5), promised to settle at a stroke more than one pressing problem of territorial division. North of the Ohio River, the meridian would have bounded a state fittingly called "Washington" in Jefferson's original version of the ordinance (Fig. 4A).[2] The state was intended for settlement by veterans of the Revolutionary War. General Washington himself had forwarded to Congress the previous June the Newburgh Petition, a document signed by nearly three hundred officers, requesting that the promise of bounty lands for soldiers be honored by a grant of land immediately west of Pennsylvania, extending from the Ohio River northward to Lake Erie, and that Congress "assign and mark it out as a Tract or Territory suitable to form a distinct Government (or Colony of the United States) in time to be admitted

---

[1] A similar problem was posed by the necessity for fixing the western end of Pennsylvania's southern boundary at a point five degrees of longitude west of a certain point on the Delaware River, as earlier noted. The boundary commissioners of 1784 met the problem by setting up an observatory near each end of the boundary. With English almanacs in hand, the commissioners at the two observatories observed the same celestial phenomena, calculating for each observation the difference between the local time of the event observed and the local time of the same event at Greenwich, England, as entered in the almanacs. Commissioners at the western observatory found the difference between their local time and that at Greenwich to be about five hours and twenty-two minutes, and commissioners at the eastern end found a time-difference of about five hours and two minutes. The exact mean difference in reckonings between the two observatories, based on two months of observing, was twenty minutes, one and one-eighth seconds. Since five degrees of longitude equals twenty minutes of time, the commissioners at the western end of the line found themselves close enough to the desired point to complete their job by linear measurement on the ground. (Thomas Hutchins Papers, Vol. III, Historical Society of Pennsylvania, Philadelphia.) Jefferson, when Governor of Virginia, had suggested this procedure. (Jefferson to Joseph Reed, In Council, April 17, 1781, in Pennsylvania Archives, First Series [12 vols.; Philadelphia, 1852-1856], IX, 78-79.)

[2] "That [state] between this [meridian] and Pensylvania and extending from the Ohio to Lake Erie, shall be called Washington." Jrnls. Cort. Cong., XXVI, 120.

one of the confederated States of America."[1] Said Washington of the land designated, "This is the tract which, from local position and peculiar advantages, ought to be first settled in preference to any other whatever."[2] In both the petition and a letter from Rufus Putnam accompanying it, the western limit of the proposed state was set beyond Jefferson's meridian, along the Scioto River (Fig. 5).[3] Timothy Pickering, in an elaborate, related scheme, known as the "Army Plan" for western settlement, proposed a "meridian line drawn thirty miles west of the mouth of the River Scioto."[4] To accommodate these army requests for a specific territory was only proper, and especially so because of the delays and disappointments already suffered by the petitioners, but to reduce the extent of this prospective military state was not inconsistent with the principles of Jefferson, who was strenuously opposed to the perpetuation in American society of distinctions based on military service.[5]

Jefferson, on the map of 1786 drawn for his Notes on Virginia, showed the Kanawha meridian as an accepted boundary line.[6] North of the Ohio it bounded "A New State," and south of the Ohio it divided "Kentuckey" from Virginia. Here, south of the Ohio, was the second pressing problem of territorial division which the meridian was intended to solve. Even while Jefferson's committee was drawing up its report, in February, 1784, a petition arrived from the inhabitants of Kentucky requesting severance

---

[1] William P. Cutler and Julia P. Cutler, Life, Journals and Correspondence of Rev. Manasseh Cutler, LL.D. (2 vols.; Cincinnati: Robert Clarke & Co., 1888), I, 159.

[2] George Washington to President of Congress, Army Headquarters, Newburgh, New York, June 17, 1783, in John C. Fitzpatrick (ed.), The Writings of George Washington from the Original Manuscript Sources, 1745-1799 (39 vols.; Washington: Government Printing Office, 1931-1944), XXVII, 17.

[3] Cutler and Cutler, op. cit., I, 159, 169.

[4] Octavius Pickering and Charles Upham, The Life of Timothy Pickering (4 vols.; Boston: Little, Brown and Company, 1867-1873), I, 546.

[5] See Jefferson's observations on the Order of the Cincinnati, in Boyd, Papers of Thomas Jefferson, X, 48-54.

[6] "A Map of the Country between Albemarle Sound, and Lake Erie . . . ," in Ford, Writings of Thomas Jefferson, III, foll. p. 84. The compilation and printing of this map is discussed in Coolie Verner, "The Maps and Plates Appearing with the Several Editions of Mr. Jefferson's 'Notes on the State of Virginia,'" Virginia Magazine of History and Biography, LIX (January, 1951), 21-33.

from Virginia by authority of Congress.[1] Jefferson's strategy, as expressed in a letter to James Madison, was simple. Accepting the separation of Kentucky as inevitable, he feared that the Kentuckians would join with settlers to the east and "Take themselves off and claim to the Alleghaney," thus removing the Kanawha basin from Virginia's jurisdiction.[2] His anxiety on this score may be judged from his plea to Madison, "For god's sake push this at the next session of assembly."[3] Ultimately, when Kentucky entered the Union, in 1792, with an irregular eastern boundary, Virginia retained even more than Jefferson's meridian would have enclosed.[4]

"We hope that N. Carolina will cede all beyond the same meridian," wrote Jefferson, in the same letter to Madison.[5] He referred to the third territorial division of immediate importance which the Kanawha meridian was expected to accomplish. A few months later, North Carolina passed an act of cession, but without regard to this line, and possibly without knowledge of its having been recommended.[6] Instead, the "extreme height" of the Great Smokey Mountains was adopted.[7] This cession was repealed at the next session of North Carolina's General Assembly, but an offer of cession made in 1789 and accepted by Congress in 1790,

---

[1] The petition appears in Boyd, *Papers of Thomas Jefferson*, VI, 552-554.

[2] Jefferson to Madison, Annapolis, February 20, 1784, ibid., p. 547. In 1783, in a draft of a constitution for Virginia, Jefferson had proposed that the General Assembly "have the power to sever from this state all of any part of it's territory Westward of the Ohio or of the mouth of the Great Kanawhay." Ibid., p. 298.

[3] Ibid. Jefferson further comments on this boundary in a letter to Madison, Annapolis, April 25, 1784, ibid., VII, 118.

[4] Compare Kentucky's eastern boundary, Fig. 4, with the course of the Kanawha meridian.

[5] Boyd, *Papers of Thomas Jefferson*, VI, 547.

[6] The North Carolina act of cession, June 2, 1784, appears in Walter Clark (ed.), *The State Records of North Carolina* (16 vols.; Goldsboro, North Carolina, 1895-1907), XXIV, 561-563. (Volume numbers continue from Colonial Records of North Carolina.) The act cites two resolutions of Congress urging state cessions, but does not mention the most recent of these, passed April 29, 1784. One infers that the legislature of North Carolina was unaware of Congressional proceedings of late April, 1784, including the passage of the Ordinance of 1784.

[7] Ibid., p. 552.

adhered to the same boundary.[1]

Passing further south, the same meridian invades South Carolina and Georgia in so patently unrealistic a fashion (Fig. 4) that doubts of serious intention are immediately aroused. The fact is that while the Ordinance was in committee, Howell wrote that the Kanawha meridian was meant to terminate at the southern boundary of North Carolina, allowing South Carolina and Georgia to extend through to the next meridian, "as their Atlantic coast falls off west."[2] A copy of a sketch-map later obtained from Jefferson in Paris, by the British diplomat David Hartley,[3] makes this allowance, and we must conclude that the failure to do so in the Ordinance represented an oversight.

The second of Jefferson's meridians (Figs. 4, 5), it will be recalled, was proposed as an Indian cession boundary as well by Jefferson's committee on Indian affairs. It was to be fixed in location by the lower end of the rapids of the Ohio River, that is, a point just below Louisville, Kentucky. The rapids, generally known in this period as the Falls of the Ohio, were looked upon by Washington as a strategic place for fortification, as was the mouth of the Kanawha River.[4] In 1780, Thomas Paine had urged the formation of a state to be bounded on the north by the Ohio River, and on the west by a meridian passing south from the rapids to the latitude of Virginia's southern boundary.[5] The same meridian, in the Ordinance of 1784, served to divide the Ohio River, below the mouth of the Kanawha River, into two

---

[1] The second act of cession was more detailed in its specification of the boundary. (Ibid., XXVI, 4.) The conflicting motives affecting North Carolina's cession and its reconsideration are examined in St. George L. Sioussat, "The North Carolina Cession of 1784 in Its Federal Aspects," Proceedings of the Mississippi Valley Historical Association, II (1908-1909), 35-62.

[2] David Howell to Jonathan Arnold, Annapolis, February 21, 1784, in Staples, op. cit., p. 479.

[3] This map, reproduced in Boyd, Papers of Thomas Jefferson, VI, 593, was obtained from Jefferson, in Boyd's opinion, at some time after August 6, 1784. (Ibid., p. 592.) It presents a curious mixture of the committee's original intention and the plan authorized by Congress, plus a deviation of the Kanawha meridian along North Carolina's western boundary that derives from neither source.

[4] Washington to the Secretary of War, Mount Vernon, June, 1785, in Fitzpatrick, Writings of George Washington, XXVIII, 168.

[5] Moncure D. Conway (ed.), The Writings of Thomas Paine (4 vols.; New York: G. P. Putnam's Sons, 1894-1896), II, 62.

nearly equal parts, while defining very closely the westward limit of settled Kentucky.[1] Further south, it split the present state of Tennessee, setting off for separate statehood a nucleus of settlement in the Nashville Basin.[2] To the north, it promised to permit the use of Lake Michigan as a natural barrier between states.

This intended employment of Lake Michigan as a natural boundary is a point which modern maps conceal, much to the damage of the reputation of Jefferson's scheme (Fig. 4). Accurate maps show the proposed meridian creating an untenable boundary slightly east of the lake, whereas Hutchins' map showed the meridian striking Lake Michigan at its southern end (Fig. 5). Accordingly, the Ordinance as first reported out of committee spoke of one state as "Westward of Lake Michigan," and of another as "Eastward thereof."[3] These clarifying remarks, being connected with the naming of states, were deleted from the Ordinance in its final form. The Hartley map of proposed states confirms the view that Lake Michigan was intended to serve as a boundary.[4]

Jefferson's plan for state boundaries was never realized. Its abandonment took one form north of the Ohio, and another south of that river. To the south, eventual boundaries were almost entirely determined by state claims and cessions.[5] Kentucky

---

[1] The meridian would have cut off Hardinsburg, as well as a settled area on the lower Salt River below Louisville, from the remainder of Kentucky. See "United States Development to 1787," map compiled by David M. Matteson, in Elroy M. Avery, A History of the United States and Its People (7 vols.; Cleveland: The Burrows Brothers, 1904-1910), VI, foll. p. 410.

[2] Ibid.

[3] Jrnls. Cont. Cong., XXVI, 119. The states, respectively, were "Michigania" and "Cherronesus." Boyd, in Papers of Thomas Jefferson, VI, 595, points out the intended use of Lake Michigan as a boundary between these two states, but he does not carry through and insist that the lake was meant to serve as a boundary all the way down to its tip.

[4] The meridian of the Falls of the Ohio strikes the southern end of Lake Michigan on this map, leaving no question as to Jefferson's conception of the boundary in this respect. (Boyd, Papers of Thomas Jefferson, VI, 593.)

[5] Soon after passage of the Ordinance of 1784, the possibility arose that eventual southern state boundaries might bear some relation to the scheme embodied in the Ordinance. In August, 1784, following North Carolina's act of cession, settlers west of the cession line, on the Holston, Wautauga and other minor streams comprising the headwaters of the Tennessee River, assembled to lay the foundation of the "State of Franklin," on the history of which

PROJECTED BOUNDARIES FOR WESTERN STATES, 1784, SHOWN ON CONTEMPORARY MAP BASE

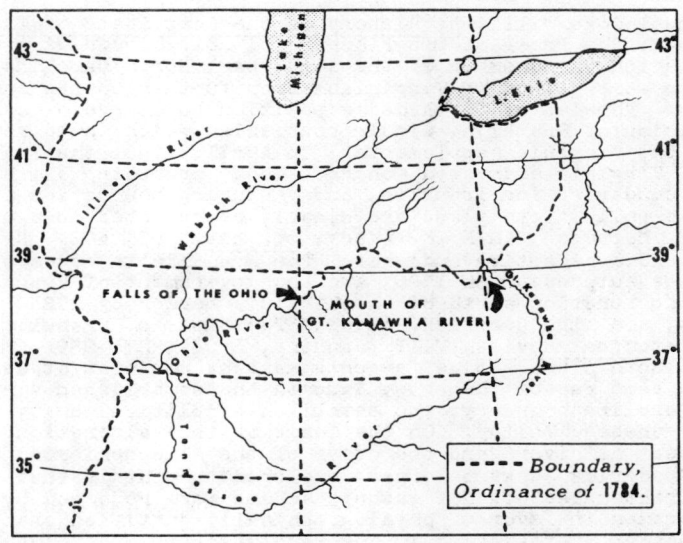

Note: Base map is Thomas Hutchins' *New Map ... Comprehending the River Ohio...* (1778), reduced from scale of 1 inch to about 20 miles to scale of 1 inch to about 160 miles.

Fig. 5

entered the Union in 1792, bounded on the east by a line which
had been authorized by Virginia as the boundary of Kentucky County, in 1776.[1] Beyond this line, Kentucky simply extended throughout Virginia's western claim south of the Ohio, to the Mississippi
River. Tennessee, entering the Union in 1796, embodied the whole
of North Carolina's claim beyond the original cession line of
1784.[2] Alabama entered the Union in 1819 with an eastern boundary
set by the Georgia cession of 1802, and a western boundary set by

---

see Samuel Cole Williams, History of the Lost State of Franklin
(New York: The Press of the Pioneers, 1933). In drawing up a
constitution in December of the same year, North Carolina frontiersmen were joined by Virginians from further up the same river
valleys. The latter forwarded a petition to Congress, asking
recognition of Franklin, within boundaries which would have included the Virginia settlements. In April, 1785, the Franklinites of Virginia addressed Congress anew, proposing slightly revised boundaries for Franklin, and venturing boundaries for Kentucky as well. Both plans are clearly represented on a map by
David E. Matteson, in E. M. Avery, op. cit., VI, 402. Neither
plan found a receptive audience. The Franklinite movement in Virginia was suppressed in 1785, and the government of Franklin
ceased to function south of the Virginia border by 1789. On this
subject, see also George H. Alden, "The State of Franklin," American Historical Review, VIII (January, 1903), 271-289.
    Both plans set Jefferson's Kanawha meridian at defiance,
for the same reason that they ignored the established Virginia-North Carolina boundary: to establish a political unity in the
upper Tennessee Valley. On the basis of this alteration, and the
employment of rivers and the crest of the Alleghenies as boundaries, Frederick Jackson Turner developed the thesis that the
"rigid rectangles" of Jefferson's scheme were rejected by the
frontiersmen in favor of physiographically justified states, in
his "Western State-Making in the Revolutionary Era," American
Historical Review, I (October, 1895), 70-78, I (January, 1896),
251-269. This interpretation fails to take the following into
account: (1) Jefferson's thirty-seventh parallel was accepted
throughout its full original length in the first plan, and a
parallel of the frontiersmen's own choosing, the thirty-fourth,
appeared in both plans; (2) the Falls-of-the-Ohio meridian was
accepted in the first plan, and simply shifted a few miles westward in the second plan; and (3) the frontiersmen adopted the
idea of splitting both present-day Kentucky and Tennessee into
eastern and western halves. One may readily discern in both of
the plans for a greater Franklin a state-making design based upon Jefferson's scheme.

[1] The line passes up Great Sandy Creek to Laurel Ridge,
and along that ridge to the southern boundary of the state. For
establishment of this line as a county boundary, see Hening, Virginia Statutes, IX, 257. For Virginia's consent to adoption of
the line as a state boundary, ibid., XIII, 17.

[2] This cession line, generally along the crest of the
Great Smoky Mountains, was described in detail in North Carolina's second act of cession, as earlier noted. (Clark, State
Records of North Carolina, XXVI, 4.)

the terms of Mississippi's earlier admission as a state.[1] Mississippi, entering the Union in 1817, represented the western remnant of Georgia's erstwhile claim.

North of the Ohio, Jefferson's scheme first met resistance in the form of a limited alternative plan favored by George Washington, and later suffered radical revision in the Northwest Ordinance of 1787. In October, 1783, it will be recalled, a Congressional committee reported on Indian policy, recommending an Indian boundary extending from the mouth of the Great Miami River to the Maumee River and Lake Erie (Fig. 3). The committee report was based in part on the advice of George Washington.[2] Both in the report and in a letter from Washington, which lay behind it, the line of demarcation was referred to as a state boundary.[3] To prevent confusion, it should perhaps be pointed out that this state, regarded as being open to general settlement, differed from the veterans' state which Washington had endorsed the previous June. In this later case, Washington was advising Congress primarily on the negotiation of peace with the Indians. "At first view," he wrote, "it may seem a little extraneous . . . that I should go into the formation of New States; but the Settlemt. of the Western Country and making a Peace with the Indians are so analogous that there can be no definition of the one without involving considerations of the other."[4]

---

[1] The western boundary of Georgia was fixed by Georgia's act of cession along the Chattahoochee River, from its mouth to a certain point in "the great Bend thereof," and thence in a direct line toward "Nickajack on the Tennessee River." Clarence E. Carter (ed.), The Territorial Papers of the United States, Vol. V: The Territory of Mississippi, 1798-1817 (Washington: Government Printing Office, 1937), pp. 142-143. The issue of whether or not Mississippi Territory should be divided into two states by the present Mississippi-Alabama border was strongly contended prior to Mississippi's admission to the Union. See ibid., pp. 332-333, 339-341, 484-487, 507-510, 731-735.

[2] For acknowledgment of Washington's advice, see Jrnls. Cont. Cong., XXV, 681. Washington's written suggestions may be found in Washington to James Duane, Rocky Hill [New Jersey], September 7, 1783, in Fitzpatrick, Writings of George Washington, XXVII, 133-140.

[3] Washington conceived of the area bounded by the proposed Indian cession line, the Ohio River, Lake Erie and Pennsylvania, as a "district for a State." But for the fact that this comprised a more compact area for a single state government, he would have preferred a cession line which would have included Detroit. (Ibid., p. 139.) The report of the committee on Indian affairs repeated Washington's observations on this subject almost exactly. (Jrnls. Cont. Cong., XXV, 691.)

[4] Washington to James Duane, Rocky Hill [New Jersey],

Washington held firmly to this plan of 1783. Writing to the President of Congress after the passage of the Ordinance of 1784, he ignored the states outlined in the Act, and referred instead to the "competent District of Land" which he had earlier envisioned.[1] Later he referred to it as "a compact State,"[2] despite the fact that its western boundary lay further west than the boundaries proposed by Putnam and Pickering, not to mention the Kanawha meridian of Jefferson's plan. In due course, Washington's basic objection to the states of the Ordinance of 1784 emerged: they would encourage widely scattered settlement. "Compact and progressive Seating will give strength to the Union," he maintained. "Sparse settlement in several new States, he wrote, "will have the direct contrary effects."[3] Sharing the general belief that ten new states had been outlined in the Ordinance, Washington referred to them as "the decies,"[4] and regarded them with misgiving as conducive to simultaneous attempts at settlement throughout the territory, to the detriment of law, good government, and the effective extension of federal aid.[5]

The great strength of Washington's policy lay in its realistic association of state-making with Indian cessions. When at last, in 1795, after the Battle of Fallen Timbers, a western line of demarcation could be dictated to the Indians which

---

September 7, 1783, in Fitzpatrick, *Writings of George Washington*, XXVII, 139.

[1] Washington to President of Congress, Mount Vernon, December 14, 1784, *ibid.*, XXVIII, 10.

[2] Washington to Arthur Lee, Mount Vernon, March 15, 1785, *ibid.*, p. 106. There are similar references in earlier letters cited here.

[3] Washington to Hugh Williamson, Mount Vernon, March 15, 1785, *ibid.*, p. 108.

[4] The widespread misunderstanding as to the number of states provided for in the Ordinance of 1784 probably was based on the fact that Jefferson had proposed ten names for western states (Fig. 4A). The ordinance as first submitted by Jefferson, erroneously represented as an act of Congress, was published in a Philadelphia newspaper, April, 1784, with an introduction which spoke of "the formation of ten new states." This impression has persisted to the present day. For discussion of this subject, with citations of erroneous graphic representations of the states, see Boyd, *Papers of Thomas Jefferson*, VI, 591-594.

[5] Washington to President of Congress, Mount Vernon, March 15, 1785, in Fitzpatrick, *Writings of George Washington*, XXVIII, 108-109.

approximated the boundary put forward by Congressional resolution in October, 1783 (Fig. 3), settlement behind the line was well advanced. Ohio, with boundaries substantially satisfying the requirements of Washington's "compact State," was admitted to the Union in 1803.[1]

A different approach to state-making in the Northwest was initiated by another Virginian, James Monroe, in 1786. With newly acquired knowledge of the Northwest, Monroe took the leadership in questioning the wisdom of the Ordinance of 1784. Writing to Jefferson, who had left Congress in May, 1784, to take up the duties of an American emissary in France,[2] Monroe said that he had returned from a trip to the Northwest "with a conviction of the impolicy of our measures respecting it."[3]

Frankly admitting that he thought the new West would be rendered too strong in Congress by the number of states allowed by Jefferson's Ordinance, he proposed to "weaken it . . . (I mean by reducing the number of the States). . . ."[4] Responding, Jefferson called Monroe's desire to divide the inland territories for the advantage of the seaboard states "a question which good faith forbids us to receive into discussion."[5] Monroe further argued, however, that to lessen the number of states by enlarging them would at the same time "render them substantial service."[6] He

---

[1] Each of the boundaries of Ohio is briefly considered in Final Report, Ohio Cooperative Topographic Survey, Vol. IV: C. E. Sherman, Miscellaneous Data (Press of the Ohio State Reformatory, 1933), pp. 17-44. A protracted dispute over the Ohio-Michigan is covered by three signed contributions, ibid., Vol. I: C. E. Sherman (ed.), The Ohio-Michigan Boundary (Press of the Ohio State Reformatory, 1916), pp. 59-115.

[2] Jefferson was elected May 7, 1784, to join John Adams and Benjamin Franklin in France for the purpose of negotiating treaties of commerce. (Jrnls. Cont. Cong., XXVI, 356.) He was elected "Minister plenipotentiary . . . at the Court of Versailles," March 10, 1785, replacing Franklin. (Ibid., XXVIII, 134.)

[3] Monroe to Jefferson, New York, January 19, 1786, in Stanislaus M. Hamilton (ed.), The Writings of James Monroe, Including a Collection of His Public and Private Papers and Correspondence Now for the First Time Printed (7 vols.; New York: G. P. Putnam's Sons, 1898-1903), I, 117.

[4] Ibid.

[5] Jefferson to Monroe, Paris, July 9, 1786, in Boyd, Papers of Thomas Jefferson, X, 112.

[6] Monroe to Jefferson, New York, January 19, 1786, in Hamilton, Writings of James Monroe, I, 117.

assured Jefferson that a great part of the Northwest was "miserably poor," pointing particularly to the parts near lakes Michigan and Erie, and to the lands along the Mississippi and Illinois rivers.[1] Doubtless he had in mind the extensive swamps and marshes of the former area in mind. As to the latter, he was referring to prairies, calling them "extensive plains wh. have not had from appearances and will not have, a single bush on them, for ages."[2] This was a pardonable, because widespread, belief about the fertility of the prairies. Monroe expected, by joining the parts of the territory which were looked upon as ill-favored with those more attractive to settlers, to hasten their inclusion in the Union.

A motion by Monroe for the reconsideration of state boundaries in the Northwest brought about the appointment of a committee whose report doomed Jefferson's plan for the boundaries of western states.[3] The committee declared:

> Such a division of the western Country cannot in the opinion of the Committee, be, in any degree practicable, conformable to the Natural boundaries of it, or for the interest of the Confederacy; according to this plan some States must be so situated as to have no advantages of Navigation; some inconveniently divided by rivers, lakes and mountains, and many of them must probably contain a large proportion of barren and unimprovable lands. . . . [I]f that Country be divided into States agreeable to the system at present adopted, the probability is that many of them will not soon, if ever, have a sufficient number of Inhabitants to form a government; the consequence of which must be, that they will continue without laws, and without order among them. . . .[4]

The committee thereupon advanced a resolution for the repeal of that part of the Ordinance of 1784 which prescribed state boundaries.[5]

Monroe's alternative plan appeared in a separate resolution by the same committee, which called upon Virginia to release Congress from the condition in her deed of cession that each state formed from the cession contain "not less than one hundred

---

[1] Ibid.   [2] Ibid.

[3] The committee, appointed to consider Monroe's motion "respecting the Cessions and division of Western lands and territory," reported March 24, 1786. (Jrnls. Cont. Cong., XXX, 131-135.)

[4] Ibid., pp. 132-133.

[5] Ibid., p. 134. Repeal became effective with the passage of the Northwest Ordinance of 1787. (Ibid., XXXII, 343.)

nor more than one hundred and fifty miles square, or as near thereto as circumstances will admit."[1] The resolution offered a substitute condition that "all the territory of the United States, lying north west of the river Ohio, shall be formed into a number of states, not less than two nor more than five."[2] Brought before Congress in July, 1786, this resolution called forth a more specific plan, moved by William Grayson, of Virginia. He proposed three lines, to accomplish the division of the Northwest: one parallel of latitude, "to touch the most southern part of lake Michigan," and two meridians, one "running due north from the western side of the Mouth of the Wabash river," and the other "running also due north from the Western side of the mouth of the big Miami."[3]

Meanwhile, Monroe had set in motion a more searching and inclusive reappraisal of the Ordinance of 1784. A committee, of which Monroe was chairman, made a report on temporary government in the Northwest in May, 1786.[4] Modified and considerably expanded by others, after Monroe's departure from Congress, this report was the basis of the Northwest Ordinance of 1787.[5] In the final drafting of this Ordinance, in July, 1787, Grayson's plan for state boundaries reappeared.[6] The fifth article of the Ordinance, covering the subject of boundaries, provided for "not less than three

---

[1] Ibid., XXX, 133-134. Massachusetts, in addition to Virginia, was at first requested to consent to elimination of this condition, since Massachusetts had supposedly ceded in conformity with the resolution of October 10, 1780, which originally stated the condition. However, only Virginia's consent was requested in the resolution finally passed, July 7, 1786. (Ibid., p. 394.)

[2] This was revised to read, "not more than five or less than three," in the resolution finally passed, July 7, 1786. (Ibid.)

[3] Ibid., pp. 390, 391. The motion was of no immediate effect.

[4] "The Plan of a Temporary Government for Such Districts as Shall Be Laid Out by the United States . . . ," May 9, 1786, in Jrnls. Cont. Cong., XXX, 252-255.

[5] The sources of the Ordinance of 1787 are brought together in the manner of an informal over-view in Theodore C. Pease, "The Ordinance of 1787," Mississippi Valley Historical Review, XXV (September, 1938), 167-180. This paper is based upon work subsequent to the publication of Jay A. Barrett's Evolution of the Ordinance of 1787 (New York: G. P. Putnam's Sons, 1891).

[6] See ordinance as "read a first time," July 11, 1787, in Jrnls. Cont. Cong., XXXII, 319-320. For final Ordinance, see Ibid., pp. 334-343, and Carter, Territorial Papers, II, 39-50.

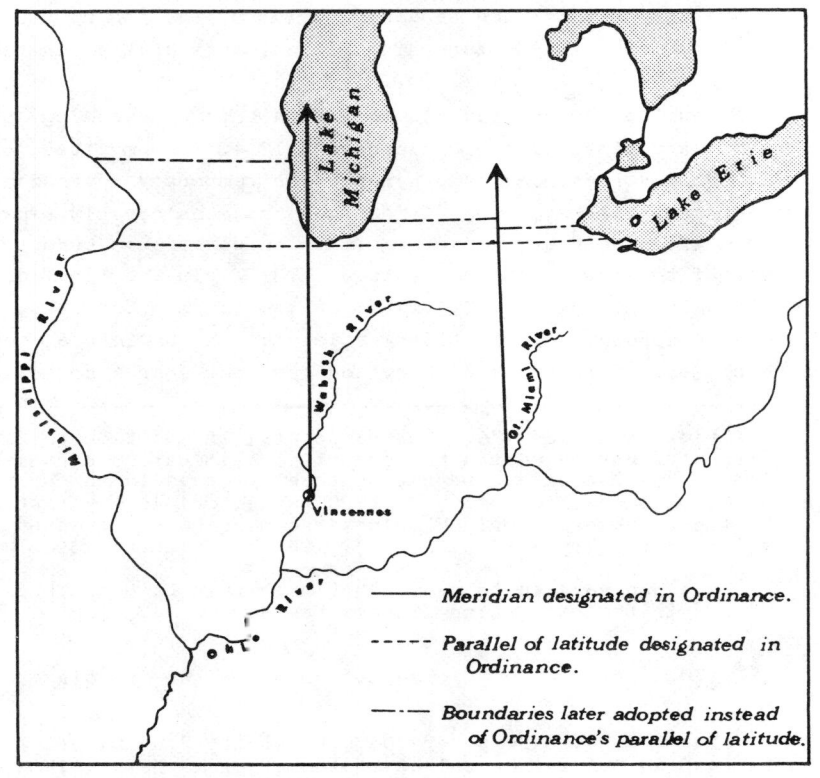

Fig. 6

nor more than five States."[1] Meridians were to run north from Vincennes, on the Wabash, and north from the mouth of the Great Miami River (Fig. 6).[2] These meridians bound present-day Indiana on the west and east, respectively.[3] Congress reserved the right to create either one or two states "north of an east and west line drawn through the southerly bend or extreme of lake Michigan" (Fig. 6).[4] Later pushed northward unequally by Ohio, Indiana and Illinois (Fig. 6), this shifted and broken line became the northern boundary of these three states. Michigan and Wisconsin, of course, were organized north of the line.

The Northwest Ordinance adopted Monroe's principle of expanded states. The boundaries prescribed in the Ordinance were Grayson's, save for the shifting of the origin of one of his meridians from the mouth of the Wabash upstream to Vincennes. Grayson's other meridian, by virtue of its adoption in the Ordinance, became the western boundary of Ohio, thus setting off Washington's "compact State," as has been indicated. Mistaken though he was in his estimate of the fertility of the prairies, Monroe was borne out in his prediction that large, sparsely populated areas could successfully be brought into the Union by their inclusion in enlarged states. Indiana, in 1816, Illinois, in 1818, Michigan, in 1837, and Wisconsin, in 1848, were admitted to statehood with extensive lands still in wilderness.[5]

Both north and south of the Ohio River, the overthrow of

---

[1] Ibid., p. 48.

[2] Ibid., pp. 48-49. Grayson's suggested westerly meridian, note, was shifted from the mouth of the Wabash to Vincennes.

[3] The meridian bounding present-day Indiana on the west, though the same as that specified in the Ordinance of 1787, terminates at a point on the Wabash River due north of Vincennes, rather than at Vincennes itself. This change, which eliminated more than one transit of the Wabash by the meridian, was written into the enabling act for Indiana's statehood, 1816. George Pence and Nellie C. Armstrong, Indiana Boundaries: Territory, State, and County (Indianapolis: Indiana Historical Bureau, 1933), p. 12.

[4] Ibid., p. 49. This differed from Grayson's motion only in making the line optional.

[5] Population distribution maps showing large vacant areas in each of these states soon after admission to the Union may be found in Henry Gannett (ed.), Statistical Atlas of the United States, U.S. Department of the Interior, Census Office (Washington: Government Printing Office, 1898), Plates 3 and 4, following p. 14.

the state outlines in the Ordinance of 1784 was complete. In
their time, however, these boundaries comprised a framework for
the anticipated operation of the first proposed national land
ordinance.

CHAPTER II

THE ORIGINAL PLAN FOR A FEDERAL
RECTANGULAR LAND SURVEY

By April 25, 1784, Jefferson's committee on public lands was ready with its proposed ordinance.[1] The committee report, entitled, "An Ordinance establishing a Land office for the United States," entirely in Thomas Jefferson's hand, was entered in the records of the Continental Congress five days later.[2] The ordinance was "read a first time" May 7,[3] and its further consideration was voted upon May 28, receiving the assent of only four of the twenty-three delegates present.[4] Put into the hands of a different committee after a new reading March 4, 1785, it was reworked into a law acceptable to Congress, the Land Ordinance of 1785.

Discussion in the present chapter will be confined to that part of the committee report of 1784 which proposed parcelling out the western lands in squares. The plan was expressed in the following words, after the preamble which declared the ordinance applicable to the territory ceded by claimant states, purchased of the Indian inhabitants, and laid off into new states:

> It shall be divided into Hundreds of ten geographical miles square, each mile containing 6086 feet and four tenths of a foot, by lines to be run and marked due North and South, and others crossing these at right angles, the first of which lines, each way, shall be at ten miles distance from one of the corners of the state within which they shall be. . . . These Hundreds shall be divided into lots of one mile square each, or 850 acres and four tenths of an acre, by lines run-

---

[1] Jefferson to James Madison, Annapolis, April 25, 1784, in Boyd, Papers of Thomas Jefferson, VII, 118.

[2] Papers Cont. Cong., XXX, 59-65. For reprint of broadside carrying proposed ordinance, see Jnrls. Cont. Cong., XXVII, 446-453. The source used for reference in the preparation of this study was Boyd, Papers of Thomas Jefferson, VII, 140-147.

[3] Jrnls. Cont. Cong., XXVI, 356.

[4] Ibid., XXVII, 453. The four favorable votes were cast by Howell of Rhode Island, Mercer of Virginia, and Williamson and Spaight of North Carolina.

ning in like manner due North and South, and others crossing these at right angles.[1]

This statement, despite its unfamiliar hundreds and its unexpected quantities, contains the essence of our national rectangular survey system.

Jefferson's ideas are strongly represented here, as will be shown in the course of this chapter, but in the interests of justice a statement made by Hugh Williamson must be recognized. In the summer of 1784, Williamson wrote from his home at Edenton, North Carolina, a few lines of praise for the recently proposed land ordinance, concluding with these words: "The general object is to oblige the Surveyors to account for the land by parallels, dotts and meridians. . . . [A]s I happen to have suggested the plan to the Committee it is more than probable that I may have parental prejudices in its favour."[2] As a former professor of mathematics and a practical astronomical observer, Williamson would have appreciated the full significance of this plan of survey.[3] In a long career as public servant, physician, local historian, and occasional contributor of papers on scientific subjects, Williamson gained a reputation for studiousness and integrity.[4] Many years after the close of the period of the Confederation, Jefferson said of Williamson, "We served together in congress, at Annapolis, during the winter of 1783 and 4; there I found him a very useful member, of an acute mind, attentive to business, and of an high degree of erudition."[5] Unless evidence to the contrary may be found, Williamson's claim should be credited.[6] Accordingly, the paragraph quoted above from the

---

[1] Boyd, Papers of Thomas Jefferson, VII, 140-141.

[2] Williamson to the Governor of North Carolina, Edenton, July 5, 1784, in Burnett, Letters of Members, VII, 564.

[3] Williamson's education and scientific activities prior to the establishment of residence in North Carolina are discussed in David Hosack, A Biographical Memoir of Hugh Williamson, M.D., LL.D. (New York, 1820), pp. 10-26. Williamson, a native of Pennsylvania, was a professor of mathematics at the College of Philadelphia, 1760-63, and was active as an astronomical observer in association with fellow members of the American Philosophical Society, in Philadelphia. (Ibid., pp. 25-26.)

[4] For details on Williamson's public services, and an indication of the scope of his published work, see ibid., pp. 61-76.

[5] Ibid., p. 67.

[6] Williamson's claim has not been entirely ignored. For

proposed land ordinance is referred to in the following pages as the Jefferson-Williamson plan. The contents of the plan are discussed below, under six heads.

The Southern System Reformed

As the time for framing legislation for the public lands approached in February, 1784, David Howell of Rhode Island foresaw a contest between two contrasting land systems for adoption into national law. The two systems, one of indiscriminate locations as practiced from Pennsylvania southward, and the other of compact survey prior to settlement as generally practiced in New England, Howell described in the following words:

> It has been the custom of the southern states to issue warrants from a land office. The person taking a warrant has to look for unlocated lands to cover with his warrant, of which he makes a return. In this way the good land is looked out and seized on first, and land of little value and of all shapes, left in the hands of the public. But this, I am told, soon rises in value, and is bought by the holders of the adjacent good lands, in their own defence. In the eastern [New England] states as you well know, the custom has been to sell a township by bonds [predetermined boundaries], or certain lots taken flush, good and bad together, and to pass out settlement in compact columns.[1]

The grid of the Jefferson-Williamson plan was later, in 1785, turned to the advantage of the advocates of township settlement, but it was originally brought forward as a means of controlling surveys under the Southern land system.

There is a belief current that the proposed land ordinance of 1784 embodied the New England principle of compact, prior survey. This interpretation has been perhaps most effectively presented by Payson J. Treat, in his study of national land dis-

---

notices of it, see Carter, Territorial Papers, II, 13, n., and Amelia C. Ford, Colonial Precedents of Our National Land System as It Existed in 1800, Bulletin of the University of Wisconsin No. 352 (Madison: University of Wisconsin, 1910), pp. 63, n., 82. In both of these instances Williamson's claim is simply cited, not supported. For examples of the customary confinement of individual attention to Jefferson, see Shosuke Sato, History of the Land Question in the United States, Johns Hopkins University Studies in Historical and Political Science, Fourth Series, Nos. 7, 8, 9 (Baltimore: Johns Hopkins University Press, 1886), p. 135; George Bancroft, History of the Formation of the Constitution of the United States of America (2 vols.; New York: D. Appleton and Company, 1882), I, 158-159, 182; and Thomas Donaldson, op. cit., p. 178.

[1] Howell to Jonathan Arnold, Annapolis, February 21, 1784, in William R. Staples, op. cit., pp. 480-481.

posal between 1785 and 1820.[1] Treat declared, following a comparison of the New England and Southern systems, that the committee report "recommended the distinctly New England system of discriminate prior surveys."[2] In an attempt to disassociate the Jefferson-Williamson plan from the New England land system, three provisions in the proposed ordinance may be brought forward. First, the ordinance preserved the Southern system of claiming lands on the authority of warrants, as will be explained more fully in the following chapter.[3] Second, surveyors were directed to proceed with the business of subdivision "beginning with the Hundreds most in demand," that is, they were allowed to scatter their surveys.[4] Third, provision was made for a warrant-holder to enter for a lot or hundred before it was laid out, simply by sufficiently identifying the lands of his choice to the district surveyor concerned.[5] All of this hardly made for prior survey in the New England sense of the term.

Although warrant-holders could claim land in advance of survey, they could receive final grants of property only in the shapes and sizes permitted by a uniform grid. By means of this innovation the proposed ordinance reformed the traditional southern system. Said Williamson, "I think the plan will prevent in-

---

[1] Payson Jackson Treat, *The National Land System, 1785-1820* (New York: E. B. Treat & Company, 1910). Treat's study, the standard work on the subject, aims "to show how the national public lands passed into private ownership during the first great period of our land system." (*Ibid.*, p. v.)

[2] *Ibid.*, p. 26. Treat, in a paper published a few years earlier, had already represented the committee report as declaring, "There shall be surveys before sales." (P. J. Treat, Origin of the National Land System under the Confederation," *American Historical Associational Annual Report for the Year 1905* [Washington: Government Printing Office, 1906], I, 236.)

[3] In Treat's view, the report "combined the New England system of surveys with the southern system of disposition." Treat, *National Land System*, p. 27.

[4] Boyd, *Papers of Thomas Jefferson*, VII, 141. By these words, a surveyor was given freedom to adopt his own order of priority among the nine hundreds comprising his district. All districts within a state, apparently, would have been thrown open to claims simultaneously.

[5] *Ibid.*, p. 143. The present author is indebted to Rudolph Freund for a recognition of this point, in his "Military Bounty Lands and the Origins of the Public Domain," *Agricultural History*, XX (January, 1946), 17.

numerable frauds and enable us to save millions."[1] Land laws of Williamson's state, North Carolina, point to the kinds of fraud which contiguous, standard lots could prevent.[2] Even more instructive is a general land law passed by Virginia in 1779, Jefferson's draft of which has recently been published.[3] In this draft we find Jefferson, as a member of Virginia's House of Delegates, framing detailed legislation for the control of indiscriminate surveys. His efforts to prohibit unduly attenuated lots, to curb surveys in excess of warranted acreage, and to prevent overlapping claims represent a step toward the basic reform contained in the later national land act.[4]

In adopting a reformed Southern system, the committee report departed from a previously established trend of Congressional policy favoring townships. A committee report of 1781 on state cessions,[5] and both the "Army Plan" and the "Financier's Plan" of previous mention,[6] anticipated township settlement in the western

---

[1] Williamson to the Governor of North Carolina, Edenton, July 5, 1784, in Burnett, Letters of Members, VII, 563.

[2] Laws for the period 1715-1783 may be found in Clark, State Records of North Carolina, Vols. XXIII and XXIV. A regulation of 1715, for example, pertained to surveys believed to be in excess of acreage granted (ibid., XXIII, 36), and a law of 1784 stipulated that "every survey shall be on the lands entered" (ibid., XXIV, 566). Williamson later touched upon the subject of land frauds in a history of his adopted state. See Hugh Williamson, The History of North Carolina (2 vols.; Philadelphia: Thomas Dobson, 1812), II, 62-64, 105-108.

[3] Boyd, Papers of Thomas Jefferson, II, 139-147. The final bill may be found in Hening, Virginia Statutes, X, 50-65. As earlier noted, this was one of two land laws passed by the Virginia House of Delegates in 1779. For a discussion of the authorship and content of both bills, see Boyd, Papers of Thomas Jefferson, II, 133-138.

[4] Ibid., pp. 143-147. Each survey was directed to be "at least one third of it's length in every part" where watercourses, mountains and previously established boundaries would permit. Excess and overlap were abuses implied by several passages in the pages cited.

[5] This report, noted as "delivered in," November 3, 1781, in Jrnls. Cont. Cong., XXI, 1098, is not there reproduced. The relevant part of the report is quoted in Albion M. Dyer, "First Ownership of Ohio Lands," New England Historical and Genealogical Register, LXIV (October, 1910), 359. Dyer also reproduces Connecticut's offer of cession of October, 1780, which called for the laying out of townships. (Ibid., p. 278, n.)

[6] Both of these plans are cited in the discussion of Jefferson's plan for western states, chap. i of this study. The "Financier's Plan" was offered in a motion before Congress. The "Army

lands. While the Jefferson-Williamson plan could accommodate township planting through the granting of entire hundreds of ten miles square, it did not protect or encourage the practice. The proposed ordinance threw open all western lands to direct claim by individual lots, whereas the New England township system required that lots be assigned only through the agency of proprietors in whom title to entire townships was vested. In the expressed view of a Massachusetts delegate, to permit the direct sale of lots meant giving up "the Plan of Townships."[1]

In reforming the Southern land system, the Jefferson-Williamson plan impaired the operation of one of that system's essential features. At the heart of the system of indiscriminate locations was the warrant-holder's prerogative of excluding undesirable lands from his claim, such as "mountains unfit for cultivation"[2] and "swamps, marshes or sunken grounds."[3] Herein lies the significance, for example, of Virginia's insistence on "good land" for its Revolutionary soldiers, in the area reserved from the general cession north of the Ohio River. The reservation was justified on the basis of an uncertainty that the "quantity of good Lands" set aside to satisfy military bounties in the Kentucky Military District would prove sufficient.[4] The squares of the Jefferson-Williamson plan could not fully satisfy this discriminating Southern land hunger, since they would compel the inclusion of lands of mixed quality within any single purchase.

On the subject of impaired freedom of choice under the proposed law, a note written by Jefferson to a foreign friend has

---

Plan," of which Putnam's letter suggesting townships is here considered a part, originated outside Congress, but it expressed the desires of a special group to the satisfaction of whose interests Congress was committed.

[1] Rufus King to Timothy Pickering, New York, May 8, 1785, in Pickering, Life of Timothy Pickering, I, 514.

[2] From Jefferson's bill of 1779 for opening a Virginia land office. Boyd, Papers of Thomas Jefferson, II, 143.

[3] The same land act, in its final form, designated such lands as second-choice, by implication. (Hening, Virginia Statutes, X, 62.) "Marsh, swamps or sunken land" appear in a comparable context in a North Carolina law of 1715. (Clark, State Records of North Carolina, XXIII, 36.)

[4] Carter, Territorial Papers, II, 8. The significance of "good land" is pointed out in Freund, op. cit., p. 12, and in William T. Hutchinson, "The Bounty Lands of the American Revolution in Ohio" (Unpublished Ph.D. dissertation, Department of History, University of Chicago, 1927), p. 50.

illuminating implications.[1] Jefferson's unqualified equation of the procedure for acquiring land under the national ordinance with the established Southern system, in this note, suggests either that he expected the land to be irregularly subdivided for further sale after its purchase from the government, or that he thought that a relatively low sale-price would reduce to unimportance the inclusion of inferior lands in lots defined by a grid.[2] As Howell once said, "The price of the land is the chief question, after all."[3] Jefferson's idea of a fair price, apparently committed to writing only in this note, was "the third of a dollar an acre."[4] The Land Ordinance of 1785 asked exactly three times this amount, as a minimum figure.

Division into Hundreds

"It [the western country] shall be divided into Hundreds," the committee report ordered. Ancient by the time of the first English settlements in America, the hundred was a subordinate division of the shire or county in England.[5] It was adopted in America only in the colonies of Delaware, Maryland and Virginia.[6]

---

[1] Boyd, *Papers of Thomas Jefferson*, VII, 220, n. The note, written by Jefferson to G. K. van Hogendorp, apprised this visitor from Holland of Jefferson's views on the sale of western lands.

[2] "The method of sale heretofore practiced by several states and now proposed by Congress has never been defeated and cannot be defeated. The first step taken is to pay the price to the public treasurer. The purchaser thereon receives from him a warrant. . . ." Jefferson to Hogendorp, *ibid*.

[3] Howell to Jonathan Arnold, Annapolis, February 21, 1784, in Staples, *op. cit.*, p. 481.

[4] Jefferson to Hogendorp, in Boyd, *Papers of Thomas Jefferson*, VII, 221, n.

[5] On the early history of the hundred in England, see F. M. Stenton, *Anglo-Saxon England* (Oxford: Oxford University Press, 1943), pp. 295-298. The author approaches through fiscal apportionments what he calls "one of the most difficult problems of Anglo-Saxon history--the origin of the institution known as the hundred," observing that "in many parts of the midlands the assessment of each hundred approximated in a round one hundred hides," and that "the correspondence of name and assessment is made more pointed by the existence of divisions assessed at 50 or 200 hides, and described as 'half-hundreds' or 'double hundreds.'" (*Ibid.*, p. 295.) The "hide" represented a land holding which supported a peasant and his household. (*Ibid.*, p. 276.)

[6] Early in the history of plans for the settlement of New England, hundreds were contemplated by the Council for New Eng-

In the last-named state the hundred was not a widely recognized division in the 1730's, whereas the parish and the county, more or less equivalent to one another in scope, flourished.[1]

The attempt to write the hundred into national land legislation was plainly the work of Jefferson, who cherished the lifelong ambition of dividing the counties of Virginia into hundreds. After his retirement from public life, he wrote to the Governor of Virginia, "I have indeed two great measures at heart, . . . 1. That of general education, to enable every man to judge for himself what will secure or endanger his freedom. 2. To divide every county into hundreds. . . ."[2] In 1778, Jefferson drew up his celebrated "Bill for the More General Diffusion of Knowledge," which would have accomplished both of these aims.[3] The bill was not passed. Twice he wrote the hundred into Virginia acts pertaining to land titles. In one of these acts as finally passed the attempt to require the creation of new hundreds was defeated by amendment,[4] and in the other the term "hundred" was struck out.[5] As late as 1810 Jefferson suggested that Virginia's

---

land, and by Sir Ferdinando Gorges, for his jurisdiction in Maine. For a discussion of the hundred in the American colonies, see George E. Howard, An Introduction to the Local Constitutional History of the United States, Johns Hopkins University Studies in Historical and Political Science, Extra Vol. IV (Baltimore: Johns Hopkins University Press, 1889), pp. 272-286.

[1]"The state . . . is formed into parishes, many of which are commensurate with the counties; but sometimes a county comprehends more than one parish, and sometimes a parish more than one county." Thomas Jefferson, Notes on the State of Virginia, ed. William Peden (Chapel Hill: University of North Carolina Press, 1955), p. 108.

[2]Jefferson to the President of the United States, Monticello, May 13, 1810, in Andrew A. Lipscomb and Albert E. Bergh (eds.), The Writings of Thomas Jefferson (20 vols.; Washington: The Thomas Jefferson Memorial Association, 1903), XII, 393.

[3]For text of this bill, which provided for common schools, see Boyd, Papers of Thomas Jefferson, II, 526-533. The aldermen of each county, in forming school districts, were "to divide their said county into hundreds, bounding the same by watercourses, mountains, or limits. . . ." (Ibid., p. 527.) Jefferson expected these hundreds to be "five of six miles square." (Jefferson, Notes on Virginia, p. 146.)

[4]In his land law of 1779, Jefferson required the certificate of any new survey to signify "the hundred wherein it lies." (Boyd, Papers of Thomas Jefferson, II, 143.) To this, the final act added, "where hundreds are established in the county wherein it lies." (Hening, Virginia Statutes, X, 57.)

[5]In Jefferson's version of a special bill concerning dis-

militia districts "be declared hundreds for the present."[1]

On each of these occasions, Jefferson was trying to establish the hundred for a single purpose--for schools, the location of property, or the organization of militia--with a view to eventual concentration of a full array of local governmental functions in the unit. Tireless in his dedication to this idea, he wrote in a letter in 1816 that the inhabitants of each hundred should have a justice, a constable, a militia company, a school, and the care of their own poor and their own roads, and that each hundred should serve as an election district. The hundreds, "of such size that every citizen can attend, when called on, and act in person," were to be, in a word, like the "townships of New England."[2]

The fact that Jefferson did not employ the term "township" in the proposed ordinance of 1784 is consistent with other evidence that he was at that time projecting views developed against a Virginia background outward upon the national domain.[3] Yet unless care is exercised, one easily assumes that Jefferson really meant "township" when he said "hundred." This is because he invested the term "hundred," in the proposed land ordinance, with a quantitative meaning which without further investigation would seem to justify in itself his use of the term as a clever substitute for the well-known New England civil division.

The hundred, as described in the proposed land ordinance, was to contain one hundred square miles (Fig. 7). By restoring quantitative significance to a term which had lost any such meaning, Jefferson was strengthened in an attempt to induce the

---

puted land titles, land holdings were to be located by hundred. (Boyd, Papers of Thomas Jefferson, II, 617.) This requirement was omitted from the final act. (Hening, Virginia Statutes, XII, 346.)

[1] Jefferson to the Governor of Virginia, Monticello, May 26, 1810, in Lipscomb and Bergh, Writings of Thomas Jefferson, XII, 393.

[2] Jefferson to Samuel Kircheval, Monticello, July 12, 1816, ibid., XV, 37-38. At this late date, Jefferson used the generic term ward to cover both township and hundred.

[3] In a well-known statement, Jefferson called the township "this most admirable of human contrivances in government." (Jefferson to the Governor of Virginia, Monticello, April 2, 1816, ibid., XIV, 454.) For use of this statement in support of the view that Jefferson had townships in mind in the 1780's, see W. A. Truesdell, "The Rectangular System of Surveying," Journal of the Association of Engineering Societies, XLI (November, 1908), 210.

general acceptance of the decimal principle into American life. Jefferson's advocacy of decimal arithmetic led almost simultaneously to his proposal for a decimal division of American coinage and a comparable division of American lands. The parallelism has been pointed out by two writers on the national land system,[1] and must have been obvious to members of the Continental Congress. The hundreds, with their division into one hundred equal parts (square-mile lots) were introduced to Congress on May 7, 1784, and made their appearance in the printed journals of Congress at some time after May 28.[2] The dollar, divided into tenths and hundredths, was put forward by Jefferson in a privately circulated set of notes on coinage which were known to some members of Congress as early as April, 1784, and appeared before Congress at large in printed form in May, 1785.[3] Jefferson's scheme for coinage was adopted by Congress, August 8, 1786.[4]

The hundred, in conclusion, represented an attempt both to stimulate the establishment of an institution of local government and to introduce decimal division into the apportionment of land. Beyond this, Jefferson had in mind a reform for which the hundreds would have served as an entering wedge. He intended to alter the entire system of English linear measurement. Essential to his plan was the geographical mile, next in the order of discussion.

The Geographical Mile

The geographical mile--better known as the nautical mile,[5]

---

[1] Amelia Ford, op. cit., p. 65, and W. A. Truesdell, op. cit., p. 211.

[2] The proposed ordinance, in Williamson's words, was "put in the Journals that the public may consider of it before the next meeting." Williamson to the Governor of North Carolina, Edenton, July 5, 1784, in Burnett, Letters of Members, VII, 563.

[3] A detailed editorial note on Jefferson's "Notes on Coinage" appears in Boyd, Papers of Thomas Jefferson, VII, 150-160.

[4] Jrnls. Cont. Cong., XXXI, 503-504.

[5] The nautical mile, the length of a minute of latitude, is about fifteen per cent longer than the statute mile. The value of the modern International nautical mile is 6076.1 feet. This value was adopted in lieu of the U.S. nautical mile (6080.2 feet), by the U.S. Departments of Defense and Commerce, in 1954. (U.S. Department of Commerce, Units of Weight and Measure: Definitions and Tables of Equivalents, National Bureau of Standards Misc. Pub. 214 [Washington: Government Printing Office, 1955], p. 4.)

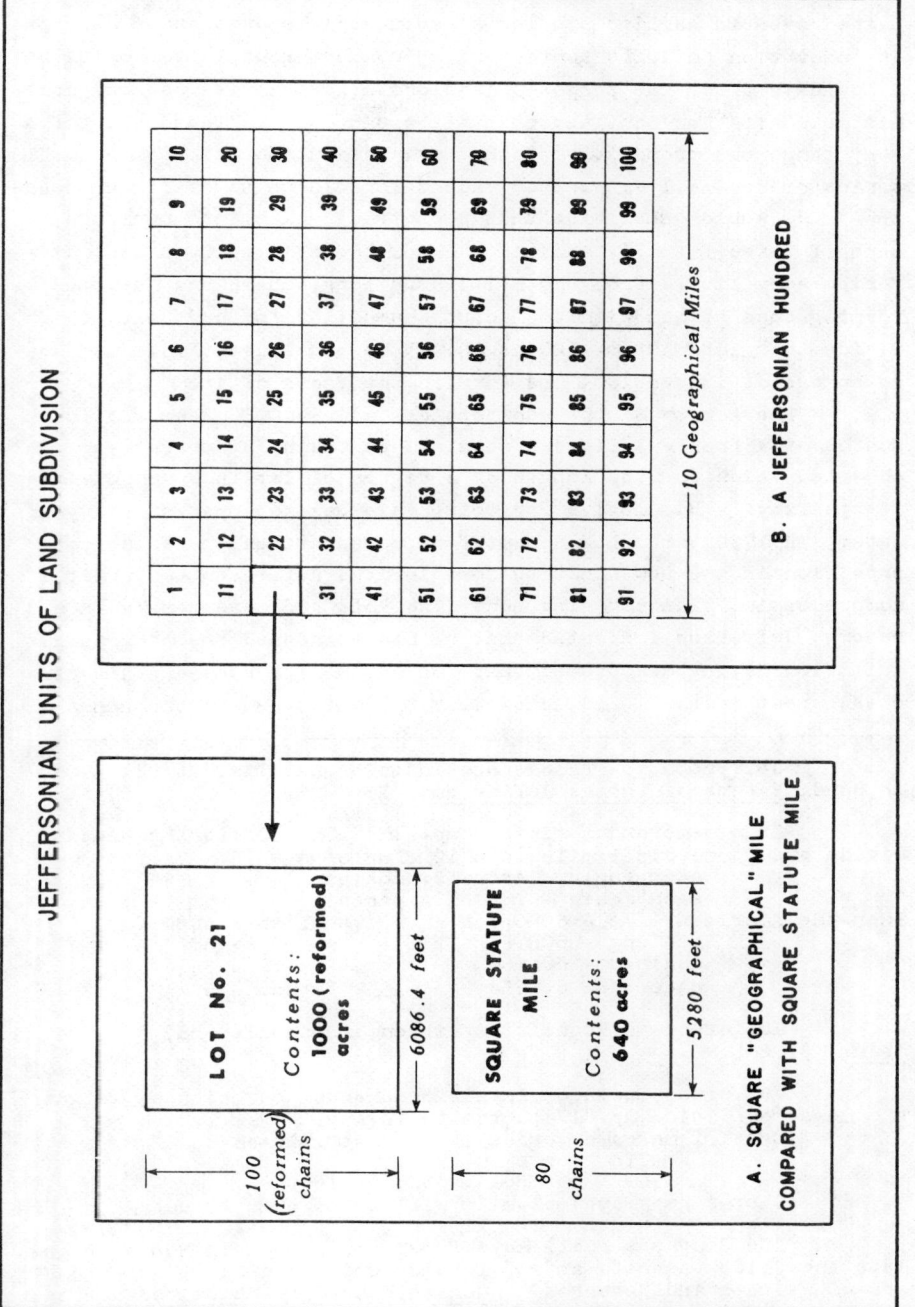

Fig. 7

and sometimes called the geometrical mile by Jefferson--has been passed over in earlier studies as an unexplained separate attempt at innovation on Jefferson's part. In a humorous letter written a few days after the proposed land ordinance had issued from committee, Jefferson confided that "it would lay the foundation of a very dangerous proposition," that is, the introduction of a new series of decimally graduated values for old units of linear measure.[1] He would "subdivide this geometrical mile into furlongs, each of these into 10. chains, each of these into 10. paces, differing very little from the British furlong, chain and fathom."[2] Turning then to his wish to "adopt the dollar for our Unit, to divide that into 10ths. 100ths. &c.," he concluded, "This is surely an age of innovation, and America the focus of it!"[3]

The length of the geographical mile--"6086 feet and four tenths of a foot"--Jefferson obtained by taking a currently accepted value for the length of a degree of latitude and dividing it by sixty.[4] His motive for doing this became apparent six years later, in 1790, when the House of Representatives, pursuant to a provision in the new Constitution, invited Jefferson to submit a comprehensive plan covering money, weights and measures.[5] In his report, Jefferson indicated that he had abandoned the geographical mile, having discovered that "no one of its [the earth's circles, great or small, is accessible to admeasurement through all

---

[1] Jefferson to Francis Hopkinson, Annapolis, May 3, 1784, in Boyd, Papers of Thomas Jefferson, VII, 205.

[2] In a memorandum (ibid., pp. 173-175), Jefferson wrote:
Divide the geometrical mile into 10. furlongs
    each furlong into 10. chains
    each chain   10. paces
Then the American mile = 6086.4 f.   English = 5280 f.
    furlong = 608.64 f.   = 660
    chain   = 60.864   = 66
    pace   = 6.0864   fathom = 6

[3] Jefferson to Francis Hopkinson, Annapolis, May 3, 1784, ibid., p. 205.

[4] Jefferson included in his memorandum on coinage, weights and measures (ibid. p. 174), the following:
Cassini makes a degree in a great mile contain
       miles   D
       69   864 = 365,184 feet
Then a geographical mile will be of 6086.4 feet.

[5] "The Congress shall have Power . . . to coin Money, regulate the Value thereof, and of foreign Coin, and fix the Standard of Weights and Measures." Constitution of the United States, Art. 1, sec. 8, par. 5.

of its parts; and the various trials to measure various portions of them, have been of such various result, as to shew there is no dependence on that operation for certainty."[1] In expressing his disappointment in the geographical mile, Jefferson revealed that in this unit he had hoped "the globe of the earth itself . . . would furnish an invariable measure."[2]

In this report Jefferson resorted, for an alternative standard, to a freely swinging pendulum adjusted in length to beat seconds of mean time.[3] Building up a new series of lengths on this basis, he arrived at a new value for the mile, which however bore a simple decimal relationship to lesser measures, as before.[4] For a satisfactory reform of linear measurement this change was not necessary.

Under the system intended by Jefferson in 1784, the U.S. public lands would have been measured with a chain divided into 100 links, and there would have been 100 chains to the mile, 10 square chains to the acre, 1,000 acres to the square mile, and 100,000 acres to the hundred (Fig. 7). Public land surveying even today is based upon the traditional English (Gunter's) chain of 66 feet.[5] Resulting measurement partakes of the decimal system at only two points: There are 100 links to the chain, and 10 square chains to the acre. The mile is divided into 80 chains, and the square mile into 640 acres (Fig. 7).[6]

---

[1] Thomas Jefferson, Report of the Secretary of State, on the Subject of Establishing a Uniformity in the Weights, Measures and Coins of the United States, published by order of the House of Representatives (New York, 1790), p. 10. Hereafter referred to as Jefferson, Report of the Secretary of State.

[2] Ibid., p. 9. Appeal to the earth as a basic reference was not peculiar to Jefferson, of course. The meter, intended to be one ten-millionth of the earth's quadrant in length, was proposed as the basis of the metric system of measurement by a committee appointed by the French National Assembly, in 1790. ("Metric System," Encyclopaedia Britannica, 11th ed., Vol. XVIII.)

[3] Jefferson, Report of the Secretary of State, pp. 10-14. This idea had been suggested by Jean Picard, in the 17th century. ("Metric System," Encyclopaedia Britannica, 11th ed., Vol. XVIII.)

[4] Jefferson, Report of the Secretary of State, pp. 22, 40.

[5] U.S. Department of the Interior, Bureau of Land Management, Manual of Instructions for the Survey of the Public Lands of the United States, 1947 (Washington: Government Printing Office, 1947), pp. 464-465.

[6] For economy of presentation, further consideration of the chain will be reserved for discussion in chap. iii, below.

We can now see in the geographical mile a passing phase of Jefferson's thought on the subject of mensuration. As it appeared in the proposed land ordinance of 1784, the geographical mile, far from being an unrelated oddity, would have entailed a sweeping reform with an ultimate effect reaching far beyond the limits of public land surveying. At the time, Jefferson said that the geographical mile had been put into the law "in such a manner as that it cannot possibly fail of forcing it's way on the people." But he added, "I doubt whether it can be carried through."[1]

Rectangles and Meridians

It will be recalled that the proposed ordinance specified (1) that the hundreds were to be "ten geographical miles square," and (2) that they were to be bounded by lines running "due North and South, and others crossing these at right angles." Reworded, the first of the two provisions called for rectilinear subdivision, and the second, for orientation to the cardinal points of the compass. Setting aside the matter of orientation momentarily, it may be said that rectangular subdivision offered the great advantage of simplicity. It was suited to the ordinary land surveying procedures of the time, which were contemplated in the law, as will be shown in the next chapter; and it assured a standard acreage figure for subdivisions, which would simplify the marketing of land. Further, if we may trust an interpretation based upon a statement made by Jefferson on a separate occasion, this manner of subdivision had a social aspect. Jefferson's report of 1790, on weights, measures and coinage, contains a highly suggestive passage on the subject of rectilinear lines. In discussing measures of capacity (quarts, gallons, etc.) Jefferson expressed himself in favor of box-like containers with plane walls meeting at right angles, in preference to cylindrical vessels. In justifying this preference he wrote:

> Cylindrical measures have the advantage of superior strength: but square ones have the greater advantage of enabling everyone, who has a rule in his pocket, to verify their contents by measuring them.[2]

---

[1] Jefferson to Francis Hopkinson, Annapolis, May 3, 1784, in Boyd, Papers of Thomas Jefferson, VII, 205.

[2] Jefferson, Report of the Secretary of State, p. 29.

Similarly, rectilinear land boundaries put it in the power of any settler, employing the most rudimentary means of measurement, to verify the contents of his purchase. We find a principle of Jeffersonian democracy implicit here, perhaps where least expected.

Orientation to the cardinal points of the compass also promised to simplify the process of wholesale subdivision of land, and at the same time to impose a control over surveyors. This control-motive was expressed in a North Carolina law of 1777, which required that each tract surveyed under the law should be bounded by "right Lines, running East, West, North and South," and should be "an exact Square or Oblong."[1] However, in North Carolina as elsewhere where one or both of these stipulations appeared, modifications were allowed. The North Carolina law, for example, accepted the use of natural boundaries as an alternative way of establishing readily verifiable limits, and it exempted tracts adjacent to prior grants or navigable waters.[2] The national ordinance, on the other hand, allowed for no exceptions. It carried regulatory strictness even further by disallowing magnetic orientation and requiring that the surveys be run "by the true meridian."[3]

We may now turn to the conflict in the law between rectangularity and cardinal orientation. The hundreds were to be of equal width (ten geographical miles), along their northern and southern boundaries, yet they were to be bounded on the east and west by meridians (lines running due north and south), which by definition are not parallel to one another, and hence could not satisfy the condition of constant width. One might say that this conflict of requirements, repeated in the Land Ordinance of 1785 and the Land Act of 1796, is notorious. Certainly it is widely known, and commonly it is regarded as a legislative "slip" whose consequences were not appreciated until surveying had been under way for several years. This altogether untenable view requires us to believe that Jefferson and Williamson, the members of two later Congresses, and several responsible surveyors, all overlooked the fact that meridians converge as they pass northward.

---

[1] Clark, State Records of North Carolina, XXIV, 46.

[2] Ibid.

[3] Boyd, Papers of Thomas Jefferson, VII, 141. The implications of this requirement will be developed in chap. iii, below.

That Jefferson suffered this elementary lapse seems doubtful not only because of his extensive reading and well-known interest in astronomical surveys, but also because of the map he compiled to accompany his Notes on Virginia, which exhibits converging meridians.[1] Even allowing Jefferson this temporary lapse, we find the conflict pointed out in the clearest of terms by Timothy Pickering, in a letter solicited for the benefit of a member of the committee which took the land ordinance in hand for revision in 1785.[2] Pickering wrote as follows:

> Each hundred is to be ten miles square, and each mile to consist of six thousand and eighty-six feet. Yet the lines marking the eastern and western boundaries are to be true meridian lines; but meridian lines converge as you increase the latitude, and to such degree, that, if you take any meridian, say at the thirty-ninth degree of latitude, and on that parallel set off ten geographical miles (equal to sixty thousand eight hundred and sixty feet) from such meridian, and then proceed northward to the forty-first degree of latitude, and there from the same meridian set off the like number of ten geographical miles, their extremity will be about eighteen hundred feet beyond the meridian of the like extremity at the parallel of thirty-nine degrees.[3]

With these remarks before them, the committee could hardly have accepted the conflicting provisions without appreciating their significance. There remains the possibility that the problem was not noticed by early surveyors, but we find that Thomas Hutchins, just before inaugurating the public land surveys in the Fall of 1785, recognized the difficulty involved, and wrote to the President of Congress from Pittsburgh asking to be "honored with Instructions on this Matter."[4]

Dismissing the belief that the conflict between rectangularity and convergency slipped into law unnoticed, we may consider the prospect for a practical solution to the problem, at the time of its first appearance. That the essentials of a

---

[1] "A Map of the country between Albemarle Sound, and Lake Erie . . . ," in Ford, Writings of Thomas Jefferson, III, foll. p. 84.

[2] The solicitor was Elbridge Gerry, who had served on Jefferson's committee in 1784. Pickering's response, a criticism of the report of Jefferson's committee, was passed on by Gerry to Rufus King, a member of the committee on public lands of 1785. (Pickering, Life of Timothy Pickering, I, 506.)

[3] Ibid., pp. 506-507. The error due to convergence, "about eighteen hundred feet," according to Pickering, would have been seventeen hundred and forty-six feet.

[4] Thomas Hutchins to the President of Congress, Pittsburgh, September 15, 1785, Papers Cont. Cong., LX, 192.

solution were near at hand may best be appreciated through the
introduction at this point of a procedure developed in later
years to cope with the problem. This procedure, which characterizes modern federal public land surveying,[1] begins with the laying
down of a "principal meridian" and a "base line," crossing at
right angles to one another (Fig. 8). One a meridian of longitude and the other a parallel of latitude, these two master lines
serve to guide the construction of rectangular subdivisions built
outward from them into four independent quadrants. These rectangular subdivisions--townships, rather than the hundreds of the
Jefferson-Williamson plan--are permitted to have meridians for
boundaries on their eastern and western sides, which increasingly
violate the rule of constant width as they are prolonged. As a
remedy, new parallels of latitude (called "standard parallels")
are run out at intervals of twenty-four miles, and new, correctly
spaced, township lines are based upon them. The early stages of
the development of this procedure, following the Land Act of
1796, will be discussed in one of the closing chapters of this
study.

Returning to the prospects for a resolution of the conflict between rectangularity and convergency in 1784, we are
brought to realize that Jefferson's state boundaries of the Ordinance of 1784, were in effect an interlocking series of principal
meridians and base lines (Fig. 4). These boundaries, under the
law, were to be laid out prior to the initiation of land subdivision, and the lines defining hundreds were to be spaced out from
the corners of the states, just as township lines would later be
spaced out from the intersections of principal meridians and base
lines. The "boxing in" of error was also provided for, in that
state boundaries would have enclosed and isolated the hundreds of
each state. This was recognized by Pickering, who, in his test
case, confined the calculated effects of convergence by the southern and northern boundaries of a Jeffersonian state.[2]

No supplementary, corrective parallels and meridians within the states were suggested in the proposed land ordinance of
1784, but it would be unreasonable to suppose that surveying was

---

[1] The procedure is summarized in U.S. Department of the
Interior, Bureau of Land Management, Manual of Instructions, 1947,
pp. 169-172.

[2] "Saratoga" (as well as "Illinoia") had these parallels
for boundaries. See Fig. 4.

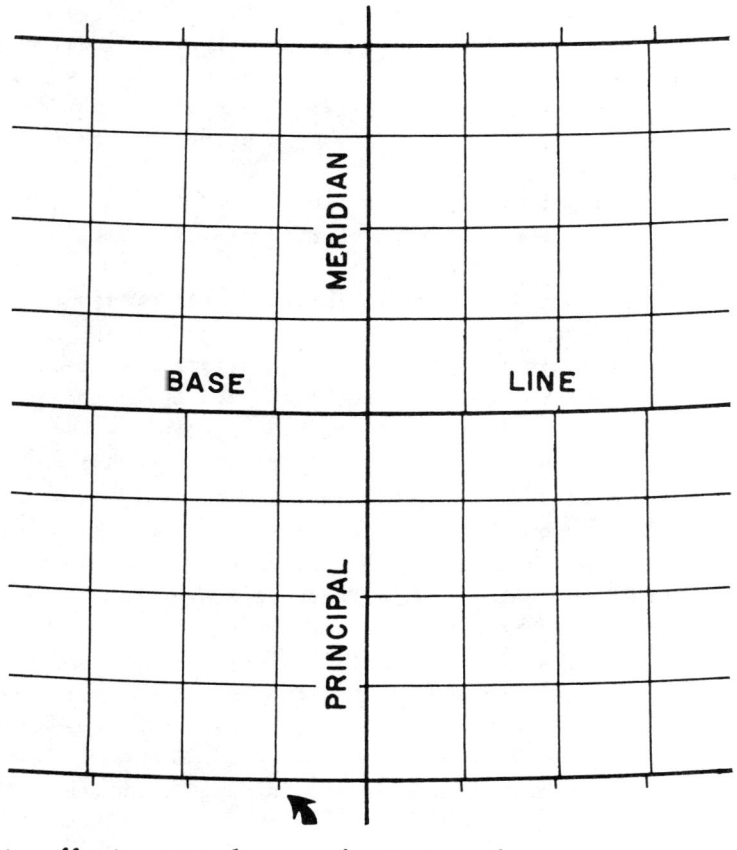

Fig. 8

expected to originate in only one corner of each state, and so allow errors to accumulate across the entire length and breadth of the state, as in Pickering's reckoning. Assuming a point of origin at each corner of a state, or at such other places as Indian cessions might allow,[1] there would have been "shatter zones" within the state where defective hundreds would have marked the collision of sets of lines of independent origin. (Such a zone lies near the eastern margin of Illinois, today.)[2] These collisions would have curbed accumulating error effectively, if not as neatly as standard parallels and guide meridians would have done.

For an adequate view of the problem of convergence as first posed, a practical consideration should be taken into account. Pickering sensibly added to his criticism, "I am aware that mathematical accuracy in actual surveys may not be expected."[3] One among many illustrations of the aptness of this remark was brought to light nearly thirty-five years later, when a corrective parallel was run eastward, ninety-six miles north of one of the earliest base lines, in southern Indiana.[4] Practically none of the township boundaries extending northward from the original base were intercepted by this parallel even approximately at points which the rule of convergency would have dictated.[5] Nearby, in Ohio, two early generally north-south lines may be found which, though parallel in intent, actually converge much more rapidly than would two true meridians.[6] Such inaccuracies point to the

---

[1] "If the Indian purchase shall not have included any one of the corners of the State, the [beginning] line shall then be run at the termination of integral miles, as measured from some one of the corners." (Jrnls. Cont. Cong., XXVII, 446.)

[2] The zone lies between Range 14 West, Second Principal Meridian, and Range 10 East, Third Principal Meridian. See map, U. S. Department of the Interior, General Land Office, State of Illinois (1911). Scale: 1 inch to 12 miles.

[3] Pickering to King [Philadelphia], March 8, 1784, in Pickering, Life of Timothy Pickering, I, 506-507.

[4] Both base line and standard parallel are discussed in chap. xi, below. See map, U. S. Department of the Interior, General Land Office, State of Indiana (1916). Scale: 1 inth to 12 miles.

[5] The intercepted township lines, counting eastward, show no regular increase in divergence as their distance from the prime meridian increases. Only the fifth and ninth lines, of the first ten, strike the parallel within even a few hundred feet of their theoretical positions.

[6] Proceeding northward from the Ohio River, the Ohio-Indiana Boundary (First Principal Meridian) and the township line

fact that surveying practices of the time comprised an independent source of distortion, sufficient to alter radically the significance of the conflicting terms in the law.

In summary, the problem raised by the convergence of meridians was appreciated at the outset of public land surveying; the eventual solution to the problem, through the use of principal meridians and base lines, was anticipated by Jefferson's state boundaries; and the contemporary importance of the problem was greatly reduced by the inability of ordinary surveying to force the issue between rectangles and meridians in a consistent fashion.

The Square-Mile Section

The hundreds, in the Jefferson-Williamson plan, were to be subdivided into "lots of one mile square each, or 850 acres and four tenths of an acre." This awkward acreage figure was needed simply for the sake of definition in the law. In the reformed Jeffersonian system of measurement each square mile would have contained 1,000 acres, as has been explained.

The words quoted above from the committee report of 1784 comprise the original proposal for the American square-mile section, of minor fame. The term "section" itself made its first appearance in the committee report of the following year.[1] It was

---

immediately to the east of it approach one another at an average rate of more than three hundred fifty feet per township in the first ninety miles or so. Two true meridians would converge at a rate of about forty feet per township. Of this glaring inaccuracy, the Surveyor General under whom the lines were run wrote as follows: "That in running North such a distance, by the needle, what we call parallel lines will converge, is a circumstance well known, but this is too much to be accounted for, on those principles, . . . it must be owing to some local cause of variation, for which I am unable to account." Rufus Putnam to Samuel Dexter, Secretary of the Treasury, Marietta, May 20, 1801, Letters Received from Surveyor General, Northwest Territory, I, 341-342, in Records of the General Land Office (Record Group 49), Interior Section, Natural Resources Records Branch, The National Archives, Washington, D.C.

[1] The report of 1785, which first brought forward both "township" and "section" as terms of public land surveying, has been particularly remembered on this account by authors associated with later public land surveying. See, for example, Frank M. Johnson, "The Rectangular System of Surveying," in U.S. Department of the Interior, General Land Office, Public Land System of the United States (Washington: Government Printing Office, 1924), p. 11; and U.S. Department of the Interior, General Land Office, Manual of Surveying Instructions for the Survey of the Public Lands of the United States and Private Land Claims (Washington: Government Printing Office, 1902), p. 5.

dropped from the 1785 land ordinance as passed into law, and reappeared in the Land Act of 1796. Thereupon the term began a career of technical usage to such effect that it has become a part of the American language with the special meaning imparted to it by public land surveying.[1]

The size of this smallest subdivision was simply that of unity in a decimal progression, as already described. In a manner of speaking, the square-mile lot played the part of penny to the hundred's dollar in a scheme for the minting of land. With regard to shape, the subject of rectangularity has been sufficiently discussed, but the subject of squareness--of equilateral rectangularity--has been deferred, and may now be disposed of in a few words. Geometric squares immediately suggest the mathematical concept of a square, that is, the product of a quantity multiplied by itself. In the mathematical operation of squaring, surveyed squares found their justification. Nothing but square forms would have permitted to the land system the "facility of Decimal Arithmetic" which Jefferson sought for, in relating lengths to areas and areas to one another.[2] The property of squareness lived on in later legislation, but it should be emphasized that it survived as a remnant of an integrated system, the original associated parts of which soon disappeared and were forgotten. The geographical mile, the hundred, and the decimal progression of reformed units of measurement were all legislative casualties, of which the square section and township remain today as reminders.

The Question of Origins

Completion, in the foregoing sections, of the discussion for which this charter was primarily intended, affords an opportunity for raising a question over which much ink has been expended: the origin of the rectangular survey system. The federal system originated, of course, in the Jefferson-Williamson plan, but it should come as no surprise to hear that many precedents for the leading features of the plan have been discovered, both in the

---

[1] *Dictionary of Americanisms*, ed. Mitford M. Mathews (Chicago: University of Chicago Press, 1951), II, 1488.

[2] "Every one knows the facility of Decimal Arithmetic . . . even Mathematical heads feel the relief of an easier substituted for a more difficult process." (Jefferson's "Notes on Coinage," in Boyd, *Papers of Thomas Jefferson*, VII, 176.) The idea of square forms may well have been borrowed from some earlier precedent. Even so, the mathematical convenience pointed out here commends itself as a probable basic reason for adoption of the idea in 1784.

form of earlier proposals and actual land divisions. The extent
to which the authors of the plan were influenced by precedents
known to them will perhaps always be open to question. Certain
precedents drawn from the laws of Virginia and North Carolina already
have been pointed out, in the belief that they were most
likely to have had a bearing on the proposed plan of federal survey.
For an idea of the number and variety of American precedents
the reader is referred to a study begun at the suggestion of Frederick
Jackson Turner, and completed as a doctoral dissertation at
the University of Wisconsin in 1908. This study, by Amelia Clewlay
Ford, is entitled, <u>Colonial Precedents of Our National Land
System as It Existed in 1800.</u>[1] It covers not only the design of
land subdivision, but policies respecting the sale and reservation
of land, and methods of public land administration. Miss
Ford reveals the type of historiography which guided her study by
speaking of the "germ" of the modern rectangular surveying system
as discovered in 17th and 18th century colonial practices.[2] Despite
the dated nature of this approach,[3] the opening chapters of
this study are rich in American precedents of possible relevance
to the federal surveying system.

Two particularly striking examples of rectilinear subdivision
are discussed and illustrated in Miss Ford's study: Sir
Robert Mountgomery's proposal of 1717 for settling his margravate
of Azilia, between the Altamaha and Savannah rivers,[4] and General
Henry Bouquet's plan for garrisoned settlements on the upper Ohio
River, published in 1765.[5] The first colonization scheme called

---

[1] Amelia C. Ford, <u>Colonial Precedents of Our National Land System as It Existed in 1800</u>, Bulletin of the University of Wisconsin No. 352 (Madison, Wisconsin: University of Wisconsin, 1910).

[2] "In the early surveys of the New England town commons and of the river lands everywhere is found the germ of the modern rectangular system." <u>Ibid.</u>, p. 18.

[3] For a brief discussion of the Teutonic germ theory and its 19th century proponents, see John H. Randall, Jr. and George Haines, "Controlling Assumptions in the Practice of American Historians," chap. ii of <u>Theory and Practice in Historical Study: A Report of the Committee on Historiography</u>, Social Science Research Council Bulletin 54 (New York: Social Science Research Council, 1946), pp. 37-40.

[4] Amelia Ford, <u>op. cit.</u>, pp. 36-37, 45. This plan, never put into effect, was designed for a settlement in South Carolina.

[5] <u>Ibid.</u>, pp. 37-38, 45-46. "Except for the early scheme

for square districts, twenty miles on a side, each comprising a symmetrically arranged unit of settlement, made up in part of square-mile cells.[1] The second proposal suggested the settlement of one hundred families in each of a series of fortified squares, a mile on a side. Each square-mile station was to have five adjoining squares of the same size at its disposal for fields, woods, and commons, the whole comprising a township.[2] Jefferson owned a copy of the book in which this second plan appeared.[3] He used information from it for his Notes on Virginia, prior to the committee report of 1784.[4] He may have been influenced by Bouquet. At least we can say that the Jefferson-Williamson plan could accommodate a scheme of frontier stations within its small, one-mile grid, just as it could accommodate New England-style townships within its large, ten-mile grid, if desired.[5]

It should be understood that American precedents at best resembled the Jefferson-Williamson plan only partially, and were in general little more than suggestive in nature. Miss Ford herself asserts that "it was a wholly new thing to use parallels and meridians for bounding . . . [land divisions] uniformly over a great area regardless of the topography of the country."[6] In

---

of Sir Robert Mountgomery, it is the only precedent which has yet been found agreeing closely with the national plan." Ibid., p. 38.

[1] An illustration of the plan is reproduced, ibid., facing p. 136.

[2] An illustration of the plan is reproduced, ibid.

[3] [William Smith], An Historical Account of the Expedition against the Ohio Indians, in the Year 1764, Under the Command of Henry Bouquet, Esq.; Colonel of Foot, and now Brigadier General in America . . . (Philadelphia, 1765). Jefferson's copy of this book is in the Jefferson Collection, Rare Books Division, Library of Congress. The plan for garrison-townships appears as a part of "Reflections on the War with the Savages of North America," pp. 51-53.

[4] Jefferson, Notes on Virginia, pp. 103-107. The borrowed material consisted of a list of estimated sizes of Indian tribes.

[5] The plan was also consistent with--and possibly partly descended from--a British proposal for land disposal of 1774. This British scheme had called for the sale at auction of previously surveyed lots varying in size from one hundred to one thousand acres. (Royal instructions to Governor William Tryon of New York, in E. B. O'Callaghan [ed.], Documents Relative to the Colonial History of the State of New York [Albany, New York: Weed, Parsons and Co., Printers, 1856-1861], VIII, 410-413.)

[6] Amelia Ford, op. cit., p. 62.

turning to Europe we encounter precedents which are distinctly more promising. Best known and most frequently pointed to are the methods of Roman surveyors. Roman surveys of cities, camps and agricultural colonies were commonly oriented by an east-west line, the _decumanus_, and a north-south line, the _cardo_.[1] Rectilinear subdivisions of the land were surveyed through the use of a simple instrument called the groma, whose horizontal crossarms, fixed at right angles to one another, provided lines of sight.[2] A standard unit in the division of agricultural land was the square _centuria_, or hundred, measuring about 2,340 feet on a side.[3]

Perhaps the earliest attempt to connect the Jefferson-Williamson plan of rectangular survey with Roman custom was made by the Honorable Joseph P. Bradley, in 1870, in an attempt to assign credit for the plan to Simeon De Witt, one-time Geographer of the United States and subsequently Surveyor General of New York.[4] The complicity of De Witt in formulating the plan is

---

[1] Edmond R. Kiely, _Surveying Instruments, Their History and Classroom Use_, National Council of Teachers of Mathematics, Nineteenth Yearbook (New York: Bureau of Publications, Teachers College, Columbia University, 1947), pp. 32, 42. Knowledge of Roman surveying is largely based upon a single collection of the writings of Roman _agrimensores_, known as the "Codex Arcerianus," a manuscript transcription dating probably from the seventh century. (_Ibid._, p. 41.) The following published work derives from this source: F. Blume, K. Lachmann and A. Rudorff, _Die Schriften der romischen Feldmesser_ (2 vols.; Berlin: George Reiner, 1848-1852). The first of these two volumes is an edited collection of the Roman texts, in Latin; the second, a commentary, in German.

[2] Kiely, _op. cit._, pp. 29-32. A groma was reconstructed by Della Corte from metallic parts discovered in Pompeii in 1912. For drawings of this reproduction, which superseded other attempts, see Kiely, _op. cit._, p. 30, and Final Report, Ohio Cooperative Topographic Survey, Vol. III: C. E. Sherman, _Original Ohio Land Subdivisions_ (Press of the Ohio State Reformatory, 1925), p. 223.

[3] E. N. Legnazzi, _Del Catasto Romano e di alcuni strumenti antichi del geodesia_ (Padua: Drucker] Tedeschi, 1887), pp. 202-203. Legnazzi calls particular attention to the Roman military colony of Lugo, at the western end of the lower Po Valley, as does Romolo de Caterini, in his "Gromatici Veteres, I tecnici erariali dell'antica Roma," _Rivista del catasto e dei servizi tecnici erariali_, II (June, 1935), 261-358. Interpretive summaries of both of these works were generously supplied by F. J. Marschner, of the U.S. Department of Agriculture.

[4] _The Centennial Celebration of Rutgers College, June 21, 1870, with an Historical Discourse Delivered by Hon. Joseph P. Bradley, and Other Addresses and Proceedings_ (Albany, New York, 1870), pp. 42-44.

highly questionable.[1] It has been suggested that his division of lands in western New York state in a fashion resembling that already authorized in national law signified that his position in the order of influence was exactly the reverse of that claimed for him by Judge Bradley.[2] In Judge Bradley's view, knowledge of Roman surveying, upon which De Witt's ideas would probably have been based, might have been gained in the course of his studies at Rutgers College, an institution "founded by men who derived their origin and traditions from Holland, a country governed by the traditions of the civil law."[3] Irrespective of the strength of his argument on behalf of De Witt, Judge Bradley was early in the field with the proposition that the American rectangular survey system was traceable to "the mode of dividing lands adopted by the Roman Agrimensores, or land surveyors."[4]

C. E. Sherman, author of one of the two best accounts to date of public land surveying in individual states,[5] was also led back to Roman surveying, in investigating the background of Col. Henry Bouquet, whose plan for the settlement of outposts, of one hundred families each, on the upper Ohio River, has been cited above.[6] Sherman wisely left open the question of whether Roman surveying "had effect through Bouquet on the American rectangular

---

[1] De Witt's only known connection with Congress at this time was as a petitioner, through correspondence, for assistance in the publication of maps which he had prepared during the Revolutionary War. See sequence of correspondence: De Witt to Washington, in T. Romeyn Beck, "Eulogium on the Life and Services of Simeon De Witt," Transactions of the Albany Institute, II (1852), 313-315; Washington to Jefferson, in Boyd, Papers of Thomas Jefferson, VII, 7-8; and Jefferson to Washington, ibid., pp. 15-16.

[2] Amelia Ford, op. cit., p. 61.

[3] Centennial Celebration of Rutgers College, p. 43.

[4] Ibid. Austin Scott, later president of Rutgers College, supported the Bradley thesis in The Targum, Rutgers College, December 12, 1884. (Sato, op. cit., p. 134.)

[5] Final Report, Ohio Cooperative Topographic Survey, Vol. III: C. E. Sherman, Ohio Land Subdivisions (Press of the Ohio State Reformatory, 1925). Hereafter referred to as Sherman, Ohio Land Subdivisions. The other thorough account of public land surveying in an individual state is John S. Dodds et al., Original Instructions Governing Public Land Surveys of Iowa: A Guide to Their Use in Resurveys of Public Lands (Ames, Iowa: Iowa Engineering Society, 1943).

[6] Sherman, Ohio Land Subdivisions, pp. 222-224.

system."[1] The essential question, after all, is that of effective influence on the committee of 1784. F. J. Marschner, a decided partisan of Roman origins, has pointed out this fact, at least: that over twenty separate issues of the writings of Roman surveyors, some of them with the subject of land division as their main theme, had appeared in printed form by 1784.[2]

Sherman, looking further for clues in Bouquet's background, turned to Holland, as had Judge Bradley, but with an eye to concrete examples of rectangular subdivision, rather than to an indefinite Roman tradition. Through correspondence and an examination of maps, he found, among other examples, the noteworthy Polder Beemster, west of the city of Edam.[3] This tract of reclaimed land, bounded on the south by the Noord Hollandsch Kanaal, is shown on modern topographic maps as an area divided by roads into perfect squares, each measuring approximately one nautical mile on a side.[4] Within each of these large squares, subdivisions, if rarely square, are regularly bounded by rectilinear canals. The canals show a distinct tendency to a spacing of one-tenth of a mile. Sherman was informed by letter that this polder was reclaimed and parcelled by the engineer Leeghwater in the first half of the seventeenth century.[5]

Hogendorp, Jefferson's "foreign friend" of previous mention, was a visitor in America from his native Holland in early 1784.[6] He may have transmitted knowledge of this type of land subdivision to Jefferson.[7] The present author, however, inclines

---

[1] Ibid., p. 224.

[2] Memorandum from F. J. Marschner, U.S. Department of Agriculture, April 15, 1956.

[3] Sherman, Ohio Land Subdivisions, p. 221.

[4] Nederlanden Topographische Dienst, Chromo-Topographische Kart des Rikjs, Sheet No. 280 (Beets) and Sheet No. 296 (Midden Beemster). Scale: 1:25,000.

[5] Sherman, Original Ohio Land Subdivisions, p. 221.

[6] Hogendorp was in Annapolis, March 26, 1784, at which time the land ordinance was under consideration by Jefferson's committee.

[7] "You have obliged me," wrote Hogendorp, "by your questions respecting the Netherlands. . . . Materials I possessed to satisfy you, but some points I had not considered in that Vieuw." Hogendorp to Jefferson [Annapolis, circa April 6, 1784], in Boyd, Papers of Thomas Jefferson, VII, 81.

to the hypothesis that Hugh Williamson, who had taken a medical degree at Utrecht[1] and was a member of two Dutch scientific societies[2]--and who, after all, stated that he suggested the rectangular plan to the committee of 1784--was the agent of transmission. Further elaborated, this hypothesis would have it that Jefferson was responsible for the idea of orientation to the cardinal points of the compass (which was not a feature of Dutch practice). If the interval of a tenth of a mile between subdividing lines was implicit in the plan as conveyed by Williamson, that idea found Jefferson predisposed to an acceptance of it, and prepared to develop and exploit it systematically. Much of the same may be said of the nautical mile. Finally, the idea of the hundred, as we have seen, was a contribution on the part of Jefferson, who had more than once tried to obtain the general adoption of it as a unit of local government in Virginia.

This hypothesis by no means rules out Roman influence. First, it assumes that the designs of Dutch engineers in subdividing reclaimed land were descended from Roman precedents in some manner which appropriate investigation could establish. Second, it allows for the expression of Roman traditions through Jefferson's contributions. The hypothesis provides a direct link with the committee of 1784 by designating Hugh Williamson as a carrier of knowledge of Dutch practices. At the same time, it leaves Jefferson's contributions unencumbered by any necessity for direct connection with the writings of Roman surveyors. The hundred, for example, had come to America from England as a territorial unit whose name had lost its quantitative significance. In restoring quantitative meaning to the term, Jefferson actually intent upon the furtherance of personal projects, as has been explained, created the illusion of direct indebtedness to Roman surveying. Again, the convenience of decimal arithmetic, though founded in Roman tradition, probably found an advocate in Jefferson more as a man of the 18th century Enlightenment than as a student of anything so specific as Roman surveying.[3] The same

---

[1] Hosack, op. cit., p. 25.

[2] The two societies were the Holland Society of Sciences, and the Society of Arts and Sciences of Utrecht. See Hugh Williamson, History of North Carolina, I, title page.

[3] Taken broadly, the entire scheme for rectangular surveying can be attributed to the "spirit of enlightenment" of the 18th century. Herbert Lehmann made this attribution in his "The Role of Law and Tradition in the Use of Agricultural Resources,"

proposition applies to Jefferson's advocacy of orientation to the cardinal points of the compass. In this connection, one should be especially wary of the apparent equivalence of the Roman <u>decumanus</u> and <u>cardo</u> and the American base line and principal meridian. Their similarity is so striking as to almost compel belief in the direct transfer of Roman surveying to the American scene. However, little more than a moment's reflection reveals their operational distinctness. The <u>decumanus</u> and <u>cardo</u> were comparatively naive in conception. As simple east-west and north-south lines, respectively, intended for highly localized use, they were quite evidently employed without regard to the geodetic problems involved in their extensive prolongation. The American base line and principal meridian--both of them state boundaries, as originally proposed--were conceived of as lines capable of indefinite prolongation: the base line as a true parallel of latitude, as distinct from the chord of a parallel of latitude, and the principal meridian as a true meridian directed poleward and not parallel to other meridians. In short, the American base line and principal meridian were designed for a spherical earth, to control and correct the very errors which the <u>cardo</u> and <u>decumanus</u>, suited to local, plane surveying only, would commit.

Until further research produces final answers, the present author would suggest that the issue of origins is likely to be more entertaining than instructive. Accordingly, in the earlier sections of this chapter attention was confined to the immediate motives and outlook of Jefferson and Williamson as legislators, in providing a rationale for their plan of survey.

As has been true of inquiries into other established ways of doing things, speculation has turned to various possible inventors of the national surveying system, as well as to precedents which might have been copied in the creation of it. In the 1880's, after a century in which the public land surveys pushed across the continent, a practicing surveyor and admirer of the system could speak of "those of us who have always lived in the West, where we know no other system."[1] Knowledge of the founding

---

Report of Seminar on Agricultural Utilization of Natural Resources, University of Chicago, Spring and Summer Quarters, 1952 (Mimeographed by Department of Geography, University of Chicago, November, 1952), pp. 2-3.

[1] H. C. Moore, "Origin and Authorship of the Present System of Government Land Surveys," <u>Journ. Assn. Engin. Soc.</u>, II (July, 1883), 283.

of the surveys had faded as admiration for them increased, and it was felt that "the memory of the founder of this system, which has proved such a benefit to our country, is, indeed, worthy of a monument"[1]--if only his name could be discovered! Three principle candidates, other than Simeon De Witt, were put forward, all of them associated with the early history of the execution of the surveys: Thomas Hutchins, Rufus Putnam, and Jared Mansfield. A vigorous claim on behalf of Hutchins (Geographer in charge of public land surveys, 1785-1789) was made in 1884, in the conviction that he had been the true author of the Bouquet plan, and that later he had been consulted on the content of the Land Ordinance of 1785.[2] Both beliefs have been shown to be unfounded in fact by the author of the best existing biography of Hutchins.[3] Putnam (Surveyor General, 1796-1803) was credited with authorship of the plan of survey, in a paper delivered in 1883, on the basis of his letter urging the settlement of Revolutionary veterans in townships.[4] While there is little doubt that Jefferson had knowledge of this letter, the plan of 1784, as we have seen, failed to adopt the traditional style of New England settlement which Putnam's letter essentially advocated. The claim on behalf of Putnam, re-affirmed in 1904,[5] was perhaps first refuted by Amelia

---

[1] Ibid., pp. 283-284.

[2] Charles Whittlesey, "Origin of the American System of Land Surveys," Journ. Assn. Engin. Soc., III (September, 1884), 275-280.

[3] Anna Margaret Quattrocchi, "Thomas Hutchins, 1730-1789" (Unpublished Ph.D. dissertation, Department of History, University of Pittsburgh, 1944), pp. 303-305. Hutchins drew the maps and diagrams for the book in which the Bouquet plan appeared. (Ibid., pp. 323-328.) Miss Quattrocchi, in denying Hutchins' authorship of the plan itself, confirms a conclusion reached in Amelia Ford, op. cit., pp. 47-53. Miss Quattrocchi's account of Hutchins' activities in 1784 and 1785 establishes his remoteness from the deliberations of Congress. (Quattrocchi, op. cit., pp. 216-231.)

[4] H. C. Moore, op. cit., pp. 282-287. Putnam's suggestion was contained in a letter cited in chap. i, above: Rufus Putnam to George Washington, New Windsor, New York, in Cutler and Cutler, Life of Manasseh Cutler, I, 171.

[5] W. A. Truesdell, "Origin of the United States Land Surveys," Journ. Assn. Engin. Soc., XXXII (April, 1904), 196. Truesdell later took the view that Putnam had been urging a customary New England procedure, rather than an invention of his own, but his belief in Putnam's influence on the plan of 1784 remained undiminished. See Truesdell, "The Rectangular System of Surveying," Journ. Assn. Engin. Soc., XLI (November, 1908), 209-214.

Ford.[1] Jared Mansfield (Surveyor General, 1803-1812) enjoyed a reputation for having originated the survey system for many years, on the authority of a popular textbook on surveying, written by his son-in-law, Charles Davies.[2] This erroneous claim was easily dismissed.[3] In seeking to bestow a just honor, well-intentioned writers, including Davies, confused in whole or part the practical role of the three patriarchs of U.S. public land surveying with the original framing of the legislation under which they acted.

In the distribution of individual honors, the names of Jefferson and Williamson must come first, followed by those of the men who re-shaped the Jefferson-Williamson plan in 1785, principally William Grayson of Virginia, and Rufus King and Timothy Pickering of Massachusetts.

Review

The principal objectives of this chapter have been the following: (1) to fix attention upon a particular paragraph in a committee report read before Congress May 7, 1784, as the foundation of our national rectangular system of land surveys, (2) to award the credit of shared authorship of the basic plan of survey to Hugh Williamson, Delegate to Congress from North Carolina, (3) to identify the proposed survey grid as an intended means of organizing and controlling surveys under the traditional Southern land system, (4) to show the full implications of such apparent oddities in the law as Jefferson's hundred and geographical mile, (5) to vindicate the judgment of early legislators respecting the convergency of meridians, and (6) to discuss the precedents for rectangular land subdivision which have been brought to light by other authors.

Toward the accomplishment of these aims, several points of interest in the report of Jefferson's committee have been

---

[1] Amelia Ford, op. cit., p. 60.

[2] Davies believed that, before the appointment of Mansfield, "lands were parcelled out without reference to any general plan." Charles Davies, Elements of Surveying and Navigation (rev. ed.; New York: A. S. Barnes & Co., 1853), p. 131. For his claims on behalf of Mansfield, see ibid., pp. iv, 131-132.

[3] Nothing more than Mansfield's belated entry on the surveying scene needed pointing out, as in Truesdell, "Origins of the United States Land Surveys," Journ. Assn. Engin. Soc., XXXII (April, 1904), 200.

slighted in the present chapter. This deficiency will be made up in the following chapter, which will be succeeded in turn by a discussion of surveying provisions in the Land Ordinance of 1785.

CHAPTER III

ADDITIONAL SURVEYING PROVISIONS IN THE FIRST
PROPOSED NATIONAL LAND ORDINANCE

The way in which the Jefferson-Williamson plan of survey was to be put into effect has yet, for the most part, to be considered. The present chapter opens with notes on the duties of registers and surveyors under the proposed ordinance. In a second section attention passes from men to means: the marked trees whereby surveyed lines were to be identified; the surveyor's chain and compass, upon which field work would rely; and the plats by which surveys were to be documented. A third section, on the numbering of surveyed lots, closes the chapter and clears the way for discussion of the Land Ordinance of 1785.

Registers and Surveyors

Responsibility for the survey and disposal of lands was to rest with a register, appointed by Congress for each prospective state, and surveyors responsible to him. The office of register, as described in the proposed law, suggests once again improvisation based on Virginia precedent. In the general Virginia land law of 1779, the Register of the Land Office figured as the principal state official concerned.[1] Whereas the example of states other than Virginia would have pointed to the creation of the office of surveyor general for western states,[2] Jefferson

---

[1] Hening, Virginia Statutes, X, 50-52, 60-61. The Virginia Register was given greater control over warrants than Jefferson had originally intended, as pointed out in editorial note in Boyd, Papers of Thomas Jefferson, II, 137.

[2] The New England states and states with a warrant system, Virginia excepted, had surveyor generals. The role of surveyor general under a warrant system appears in the following account of land patenting in Pennsylvania: "When a man had discover'd a piece of vacant land in any part of the province he applies to the Secretary for a warrant to have the same Surveye'd to him. . . . The Secretary directs his Warrant to the Surveyor General, who keeps the same in his Office, and directs a Copy thereof to the Deputy Surveyor of the District where the Land lies. When the Survey is made, the Deputy Surveyor returns the draught thereof to

quite apparently borrowed the familiar register from experience, and added to his responsibilities that of supervising surveys.[1] The register of a new western state would have combined this task of supervision with the preparation of warrants for dispatch to the Treasurer of the United States, and the granting of final deeds to land.[2]

"The Office of Register which Congress are about to establish on an Extensive Scale would undoubtedly be very acceptable to me," Thomas Hutchins, Geographer of the United States, wrote to a friend soon after the committee report was placed before Congress.[3] He was ready to resign his post in favor of the profit, in fees, which he expected from a registership. Simeon De Witt, also Geographer of the United States,[4] had confided before any federal legislation was proposed, "If a new state is to be laid off adjoining Pennsylvania and Virginia, as has been expected, I have hopes of . . . [being appointed] surveyor general to such a state. . . ."[5] The hopes of both men were fulfilled in somewhat altered form: De Witt became a surveyor general, but of New York rather

---

the Surveyor-General, where it lies till the Man applies for a Return thereof to the Secretaries Office, & by this return to the Secretary (upon the Payment of the Remainder of the purchase Money to the Receiver General), he draws out a Patent. . . ." Lewis Evans, "Brief Account of Pennsylvania," reproduced in Lawrence Henry Gipson, *Lewis Evans* (Philadelphia: The Historical Society of Pennsylvania, 1939), p. 136.

[1] Centralized coordination of surveying was notably lacking in the general land law of 1779. See Hening, *Virginia Statutes*, X, 50-63.

[2] The Treasurer was to send the warrants "to the Commissioner of the loan office for the United States in each of the states within the Union, the Treasurer countersigning them on parting therewith." (Boyd, *Papers of Thomas Jefferson*, VII, 142.) Whereas the Virginia Register made up grants for the Governor's signature, the federal registers were authorized to make grants on their own authority. (Hening, *Virginia Statutes*, X, 60-61, and Boyd, *Papers of Thomas Jefferson*, VII, 143.)

[3] Hutchins to John Montgomery, Philadelphia, May 19, 1784, John Montgomery Papers, Chicago Historical Society.

[4] Congress passed a resolution, July 11, 1781, that both De Witt, Geographer to the Main Army, and Hutchins, Geographer to the Southern Army, be styled "Geographer to the United States of America." (*Jrnls. Cont. Cong.*, XX, 738.)

[5] De Witt to Washington [January, 1784], in T. Romeyn Beck, *op. cit.*, p. 316.

than of a new western state;[1] and Hutchins was later given command of federal surveying under the Land Ordinance of 1785, but as a salaried geographer, not as a register in receipt of fees.[2]

Surveyors, appointed by Congress, were to act under bond and to be subject to suspension by their state registers "for negligence or malversation."[3] The survey of state boundaries was to precede the work of these men, who were to begin with the laying out of hundreds. Then, in the words of the law, "the Hundreds being laid off and marked, nine of these shall be assigned as a district to each surveyor, who shall then proceed to divide each Hundred of his district into lots. . . . beginning with the Hundreds most in demand."[4] These districts were probably intended to be "stalking horses" for eventual counties, just as the surveyors' hundreds hopefully adumbrated political hundreds. The surveyors, in turn, were, in effect, Virginia county surveyors transplanted to the West.

The qualifications of a Virginia county surveyor may be illustrated from the early career of George Washington. In 1749, the young Washington was commissioned Surveyor for Culpeper County, Virginia.[5] Qualifying at a time when settlement was extending rapidly up the Shenandoah Valley, he was in a position both to accumulate fees in cash for laying out the claims of others, and to scout for land for his own modest investment.[6] In the preceding year he had crossed the Blue Ridge Mountains to view this area for the first time, in the company of James Genn, a veteran surveyor, and George William Fairfax, agent for the proprietor of the lands which Washington was soon to be surveying on his own.[7] Once commissioned, Washington was authorized to survey

---

[1] De Witt was commissioned Surveyor General of New York, May 13, 1784, one week after the report of Jefferson's committee was given its first reading in Congress. (Ibid.)

[2] See section headed "Geographer and Surveyors," chap. iii, below.

[3] Boyd, Papers of Thomas Jefferson, VII, 142.

[4] Ibid., p. 142.

[5] Douglas Southall Freeman, George Washington, A Biography (4 vols.; New York: Charles Scribner's Sons, 1948-1951), I, 234.

[6] Washington was able to patent approximately one thousand acres of land in October, 1750. (Ibid., p. 243.)

[7] Ibid., pp. 202-210.

in any county in Virginia, and he was profitably occupied, so far as weather and family affairs permitted, until the Fall of 1751.[1] The abilities which his survey records reveal are these: that of taking and recording compass bearings to the nearest degree, that of designing a polygon to include desired land and satisfy a given acreage figure, that of representing a traverse on a simple plan or plat, and that of identifying trees.[2] He displayed, of course, in addition, a capacity for the conduct of business, ability to recruit and manage small survey crews, and the hardihood necessary for life in the field.[3]

Men with such qualifications were to carry the warrant system onto national lands. Under the proposed law the holder of a warrant would designate to a surveyor the lands of his choice, and receiving in due course a certificate signifying the verified bounds of his claim, would submit it with his warrant to the appropriate state Register for the issuance of a deed.[4] For their services, the surveyors were to receive fees. In this connection, an additional surveyors' attribute, of which Washington was free, should be emphasized once more: dishonesty. If Virginia or any other state with a warrant system had reviewed its land grants at this time, a situation similar to that uncovered by a committee of the General Court of Massachusetts, appointed in 1783, would doubtless have been revealed. This committee, after examining grants made outside the township system in the present state of Maine, reported as follows:

> They were laid out in irregular forms, and frequently at a distance from any located Lands. Tracts thus laid out in general, contain at least ten percent more than the quantities specified in the grants on which they are founded, and in some instances they have been found to contain nearly double the quantities intended by the original Grants.[5]

---

[1] Washington terminated this phase of his career by sailing for Barbados, September, 1751. (Ibid., p. 248.)

[2] These judgments are based upon survey records reproduced in George Washington, Journal of My Journey over the Mountains, While Surveying for Lord Fairfax, Baron of Cameron, in the Northern Neck of Virginia, Beyond the Blue Ridge, in 1747-8, ed. J. M. Toner (Albany, N.Y.: Joel Munsell's Sons, 1892), pp. 74-131.

[3] See account of Washington's surveying experiences in Freeman, op. cit., I, 213-246.

[4] Boyd, Papers of Thomas Jefferson, VI, 143.

[5] Report of the Committee for the Sale of Eastern Lands: Containing Their Accounts from the 28th of October, 1783, to the 16th of June, 1795 [Boston, 1795], p. 2.

The imposition of a regular grid by the proposed national land ordinance was intended, as Williamson made clear, to curb such irregularities. As a further protection, survey returns were to be tabled by the Register for a period during which any counter-claimant could file a "caveat" against the issuance of title to a parcel which a surveyor might have certified more than once.[1] Surveyors were suspected in advance of wishing to increase their fees by certifying overlapping claims.[2]

Finally, one measure of control was conspicuously lacking in the proposed ordinance. There was no provision for tests of competence on the part of surveyors. This was necessarily a conscious omission, since Jefferson had taken care to carry over such a provision from colonial law into Virginia's state land law. In colonial Virginia, the power of licensing surveyors was vested in the College of William and Mary,[3] and this power, together with the added responsibility of reviewing lists of surveys, was assigned to the same institution in Jefferson's draft of Virginia's general land act of 1779.[4] It is doubtful, however, that Congress could have required the states of the Union to furnish licensed surveyors, under the Articles of the Confederation, and the appointment of a federal board of review would probably have exceeded the limits of desirable federal employment capacity, as viewed by many delegates to Congress, Jefferson not least among them.[5]

---

[1] Boyd, Papers of Thomas Jefferson, VII, 144-145. The "caveat" was a normal feature of the southern land system.

[2] The law provided: "Where he shall have admitted more locations than one or the same land, he [the surveyor] shall restore the fees received from the party whose location shall be set aside." (Boyd, Papers of Thomas Jefferson, VII, 146.)

[3] Washington's commission, for example, was issued by the College of William and Mary. (Freeman, op. cit., I, 234.)

[4] Boyd, Papers of Thomas Jefferson, II, 141, 142, 144. See also act as passed, in Hening, Virginia Statutes, X, 53, 57-58.

[5] Reduction of the Civil List, "both as to the Number of Officers, and the Salaries of those that are to remain" was one of the principal concerns of Congress in the spring of 1784, in the view of Roger Sherman of Connecticut. Sherman to the Governor of Connecticut, Annapolis, March 29, 1784, in Burnett, Letters of Members, VII, 479. Jefferson was a member of a committee charged with scheduling such a reduction. The committee report, written by Jefferson, appears in Jrnls. Cont. Cong., XXVI, 125-127.

Trees, Chain, Plat and Compass

Four standard elements of land surveying appeared in the proposed land ordinance: marked trees, the chain, the plat and the compass. In providing for the employment of these means of parcelling land, intended perpetuation of established practices was the general rule. Discussion of each of the four elements follows.

Just as Virginia's land law had earlier called for lots to be "bounded plainly by marked Trees,"[1] so the federal law called for lines to be "plainly marked by chaps or marks on the trees."[2] In hindsight, this may appear to have been a way of requiring the use of "bearing trees," which later came to be a regular feature of national public land surveying. Two or more trees standing near a post or other corner marker were blazed and inscribed, and their courses and distances from a corner were recorded, to assist in any later determination of the corner's position.[3] By surveying custom in Virginia, on the other hand, the trees themselves were employed as corners.[4] Allowing, of course, for the occasional coincidence of an accurately surveyed corner with a tree, this practice put the course and length of bounding lines second to the convenience of trees in order of importance. For want of any indication to the contrary, it is assumed that the federal law contemplated continuation of this practice.[5] We find the terms necessary for the authorization of

---

[1] Hening, Virginia Statutes, X, 57.

[2] Boyd, Papers of Thomas Jefferson, VII, 141.

[3] On the use of bearing trees in U.S. public land surveying, see U.S. Department of the Interior, General Land Office, Manual of Surveying Instructions for the Survey of the Public Lands of the United States and Private Land Claims (Washington: Government Printing Office, 1902), pp. 53-54; and Lowell O. Stewart, Public Land Surveys: History, Instructions, and Methods (Ames, Iowa: Collegiate Press, Inc., 1935), pp. 120, 122, 144.

[4] The following extract from Washington's survey notes will serve to illustrate this practice: "Beginning at two white Pines and a Pitch and running thence So 62 Et Three Hundred Poles to a Chesnut, Pine and Spanish Oak on a Mountain Side thence No 28 Et Two hundred & Six poles to two white Oakes and a Hickory. . . ." (George Washington, Journal of My Journey, p. 80.)

[5] It should be added that tree-marking was not necessarily to be confined to corners. The proposed law could have contemplated the marking of trees standing in the course of a surveyed line, as well. Such trees came to be known as "line trees," in later public land surveying.

better corner-marking, for example, in the instructions for bounding a 17th century grant to a Massachusetts town: ". . . a trackt of land . . . to be wel bounded by marke trees with BT set one the barke of the trees and heapes of stones or by diging a litle square hole that their maye be markes upon every side within 50 rodes of one another."[1] In contrast, the proposed federal law could only be interpreted as limiting the marking of surveys to the use of trees, without regard to their exact coincidence with accurately surveyed positions.

A second standard element appeared in the stipulation that "lines shall be measured with a chain." This was the era of the surveyor's chain, which had superseded ropes and wooden poles as the usual means of land measurement,[2] and which would be superseded in turn by the steel tape.[3] The persistence of earlier nomenclature can lead to confusion in interpreting survey records of this period, since "poles" continued to appear as units of measure in surveys employing the chain.[4] Similarly, in modern nomenclature, "chaining" continues to be spoken of, where measurement with a steel tape is meant.[5] Of the surveyor's chain an early American book on surveying had the following to say: "The instrument most in use, for measuring the Sides of Fields

---

[1] Samuel A. Bates (ed.), *Records of the Town of Braintree, 1640-1793* (Randolph, Massachusetts, 1886), p. 12.

[2] Robert Gibson, *A Treatise of Practical Surveying; Which Is Demonstrated from Its First Principles* (5th ed.; Philadelphia, 1789), p. 129. See also Kiely, op. cit., p. 236.

[3] Steel tapes first came into use in the United States around 1860. (John L. Culley, "Steel Tapes," *Journ. Assn. Engin. Soc.*, VI [August, 1887], 306.) By 1902, use of the steel tape was officially sanctioned for U.S. public land surveying. (U.S. Department of the Interior, *Manual of Instructions* [1902], p. 22.) By 1930, the steel tape alone was authorized. (U.S. Department of the Interior, General Land Office, *Manual of Instructions for the Survey of the Public Lands of the United States, 1930* [Washington: Government Printing Office, 1931], p. 25.)

[4] Washington, for example, used the pole as his unit of measure, but the chain as his instrument of measurement. Use of the chain is made evident by his listing of chainmen in *Journal of My Journey*, pp. 78, 80, 88-130. For a contemporary chainman's oath, in Connecticut, see E. D. Kingman, "Roger Sherman, Colonial Surveyor," *Civil Engineering*, X (August, 1940), 515.

[5] This is exemplified by "Basic Problems of Chaining," chap. vii of John Clayton Tracy, *Surveying, Theory and Practice* (New York: John Wiley & Sons, 1947), pp. 69-79.

is Gunter's Chain,[1] which is in length 4 Rods or 66 feet; and is divided into 100 equal parts, called Links, each containing 7 Inches and 92 Hundredths."[2] Chains of slightly later date examined by the present author have been found to be made of iron wire, about one-tenth of an inch in diameter, forming one hundred straight segments, each segment joined to its neighbor by two rings. The length of a link, as given above, was found to include one straight segment and a ring at each end.[3] Jefferson's geographical mile, it will be recalled, would have brought in its wake a chain of new length (60.86 feet, as against 66 feet), but in design and division, it may be assumed, Gunter's chain would have remained unchanged.

A third established element of land surveying made its entry into national land law through the provision that all surveyed lines "shall be exactly described on a plat."[4] Although the "fair and true plat" of Virginia land law may have been the direct progenitor of this type of record in the national ordinance,[5] the plat (also called "plot" and "plott") was familiar throughout the seaboard states, as a simple drawing or plan of the boundaries of a property survey. Plats, whether of New England towns[6] or individual southern claims,[7] were primarily in-

---

[1]This was one of several practical inventions for facilitating measurement and calculation made by Edmund Gunter, English mathematician (1581-1626). "Gunter, Edmund," *Encyclopedia Britannica*, 14th ed., Vol. XI.

[2]Abel Flint, *A System of Geometry and Trigonometry: Together with a Treatise on Surveying* (2d ed.; Hartford, 1808), p. 35. General use of Gunter's chain in the 1780's is indicated in Gibson, op. cit., p. 129. For chains of other lengths, chains with other divisions, and chains based on nominally identical units having other values, see ibid., pp. 129-133.

[3]Chains, ranging in date from circa 1810 to circa 1850 have been examined at the Ross County Historical Society, Chillecothe, Ohio, the Wooster Museum, Wooster, Ohio, the Ohio State Archeological and Historical Society, Columbus, Ohio, and the Office of Auditor of State, Indianapolis, Indiana.

[4]Boyd, *Papers of Thomas Jefferson*, VII, 141.

[5]Hening, *Virginia Statutes*, X, 57.

[6]For examples, see plats of townships in New Hampshire and Vermont in Albert S. Batchellor (ed.), *New Hampshire Provincial and State Papers* (40 vols.; Concord: State of New Hampshire, 1867-1943), Vols. XXIV-XXVII, passim.

[7]For examples of plats of individual claims, see *Original Land Titles in Delaware Commonly Known as the Duke of York Record* (Wilmington, 1903), pp. 37-84.

tended to illustrate and support the verbal descriptions submitted by surveyors, for the security of title to land. Applied to the national domain, as a part of the preparation for its sale, their purpose was the same. Surveyors under a Virginia law drafted by Jefferson were to show on their plats streams and "other notable objects which occur in, coincide with, or are adjacent to, every line, with their distances from one another, and from such lines."[1] The aim of strengthening the identification of boundaries is evident. Similarly, the national ordinance directed that "[on the plat] shall be noted, at their proper distances, all watercourses, mountains, and other remarkable and permanent things, over and near which such lines shall pass."[2] In both passages a tendency to require description beyond the strict needs of cadastral surveying can be recognized. Building on this foundation, the Land Ordinance of 1785 carried descriptive requirements a step further, and thereby launched the public land surveys upon a career of recorded exploration in advance of settlement, as will be discussed later in this study.

Among the proposals made in the committee report of 1784 and eliminated from the ordinance of the following year was a plan for making state-wide compilations of plats. Each register was directed to procure from the surveyors "the plats of all lines, measured and marked by them in the preceeding half year, to be by him collated, and reduced into a general map of the whole state for which he acts."[3] Further, the register was to submit annually to the Secretary of Congress "a copy of such portions of the said general map as shall have been formed, or further filled up, during the preceeding year."[4] This proposal anticipated by more than half a century maps showing the progress of surveys by state and territory, submitted by surveyors general. These maps, which may be found accompanying the annual reports of the Commissioner of the General Land Office in the Congressional Documents Series,[5] exemplify the kind of graphic summarizing

---

[1] Boyd, Papers of Thomas Jefferson, II, 432.

[2] Ibid., VII, 141.   [3] Ibid., p. 142.   [4] Ibid.

[5] See U. S. Congress, Senate, Report of Commissioner of General Land Office, Senate Doc. No. 11, 25th Cong., 2d Sess. This report was the first to include maps exhibiting the progress of surveys. One map from this report of 1837, and eight maps from subsequent reports, through 1859, make up a series illustrating the advance of surveying across the State of Iowa, in Roscoe L. Lokken, Iowa Public Land Disposal (Iowa City: The State

which would have been effective from the beginning, under the ordinance of Jefferson's committee.

A fourth established element of land surveying, the magnetic compass, was recognized by the committee report in an important sentence regulating its use, which passed almost word-perfect into the Land Ordinance of 1785:

> The Surveyors shall pay due and constant attention to the variation of the magnetic meridian, and shall run and note all lines by the true meridian, certifying with every plat what was the variation at the time of running the lines thereon noted.[1]

By holding the surveyors to the true meridian, this regulation threatened to precipitate a conflict between rectangularity and convergency, as discussed in the last chapter.

The background for this regulation is apparent. Jefferson had already tried to compel the use of the true meridian in Virginia, by writing into a law (which was not passed) the requirement that each survey be represented on a "plot, protracted by the true meridian . . . [showing] the variation thereof, towards the east or west, from the magnetical meridian."[2] It should be emphasized that neither here nor later, in the national law, was abandonment of the magnetic compass proposed. Jefferson was simply requiring that a verified correction be applied to magnetic bearings, by which they would be converted to true bearings. By common reference to true north, separate surveys could be related to one another in a fashion not otherwise possible. The disagreement of lines run with different compasses, at various times, and subject to local attractions, may be readily imagined. Unsuccessful in an attempt to remedy a confused situation in Virginia, Jefferson acted to secure the national grid against a similar plight.

Use of the uncorrected needle in land surveying was usual in America at this time.[3] The instrument commonly employed for

---

Historical Society of Iowa, 1942), pp. 21, 27, 29, 33, 39, 43, 47, 49, 51.

[1]Boyd, Papers of Thomas Jefferson, VII, 141.

[2]This passage appears in "A Bill for Ascertaining the Salaries and Fees of Certain Officers," ibid., II, 432.

[3]For remarks which imply or directly state that the uncorrected needle was in general use, see statement by Simeon De Witt, quoted in T. Romeyn Beck, op. cit., p. 322; Timothy Flint, op. cit., p. 80; Robert Gibson, The Theory and Practice of

running magnetic lines was the circumferentor.[1] It was a compass measuring about six inches in diameter, graduated to give readings in degrees, fitted with sight vanes, and mounted by means of a ball and socket on a staff ("Jacob's staff") or tripod.[2] Of immediate interest is the fact that some of these circumferentors were so constructed as to allow the magnetic declination (or "variation," as expressed in the ordinance) to be set off. That is, the compass circle could be slightly rotated and fixed in a deflected position, such that corrected readings could be taken directly from the needle. Even without such an adjustment, corrections (in whole degrees) could be easily applied, of course.

It should be noted that surveyors, under the proposed ordinance, were to pay "constant attention" to the variations of the needle, and that the variations noted on the plats were to be based on readings "at the time of running the lines." The first caution probably referred to deflections caused by local attraction, but the second pointed to the author's awareness, or strong suspicion, that magnetic bearings were subject to change from other causes. Many years later, Jefferson wrote, "The law of those variations is not yet sufficiently known to satisfy us that sensible changes do not sometimes take place at small intervals of time and place."[3] In the face of these changes, how were the surveyors expected to check their needles?

Two relatively easy means of relating the magnetic bearings given by circumferentors to true bearings offered themselves: observation of the North Star, and observation of the sun at the

---

Surveying (New York, 1821), p. 359; and Commonwealth of Massachusetts Land Court, Manual of Instructions for the Survey of Lands and Preparing Plans for the Land Court (Boston, 1913), p. 29.

[1] Circumferentors are described in Gibson, Treatise of Practical Surveying, pp. 148-150, and in Newton C. Brainard, "Colonial Surveying Instruments," Connecticut Historical Society Bulletin, XIV (April, 1949), 10-12.

[2] This general description is based both upon the sources cited in the note immediately preceding, and upon personal examination of circumferentors at the Campus Martinus Museum, Marietta, Ohio, the Wooster Museum, Wooster, Ohio, the Ohio State Archeological and Historical Society, Columbus, Ohio, and the Office of the Auditor of State, Indianapolis, Indiana.

[3] Jefferson to Governor Wilson C. Nicholas, Poplar Forest, April 19, 1816, in Lipscomb and Bergh, Writings of Thomas Jefferson, XIV, 483.

time of its rising or setting. Jefferson favored ascertaining the magnetic variation at sunrise, when he later expressed himself on the subject. He wrote, "To render these observations of the variations easy, and to encourage their frequency, a copy of the table of amplitudes should be furnished to every surveyor. . . ."[1] By amplitude was meant the angular distance north or south of due east at which the sun would rise on a given date.[2] Of the alternative method, based on observation of the Pole Star, an American author wrote several years after the framing of the ordinance, "It has been adopted by many surveyors."[3]

Though comment on marked trees, the chain, the plat and the compass, as they appeared in the report of Jefferson's committee, concludes at this point, their consideration will be resumed later in this study, in connection with surveying in the field under the terms of the Land Ordinance of 1785.

The Identification of Lots

The possibility of identifying parcels of land by number was not least among the benefits conferred by the adoption of a uniform grid for land subdivision. It represented a release from the cumbersome colonial tradition of description by metes and bounds, of which the following extract from a deed, dated 1742, may serve as an example:

> Beginning at a white oak in the fork of four mile run called the long branch & running No 88° Wt three hundred thirty eight poles to the Line of Capt. Pearson, then with the line of Pearson No 34° Et One hundred Eighty-eight poles to a Gum on the So Wt side of the run corner to Pearsons red oak & chesnut land, then down the run & binding therewith So 54° Et Two hundred & ninety poles to the beginning, Containing One hundred Sixty six Acres, . . .[4]

The land conveyed by this deed lay in Fairfax County, Virginia, across the Potomac River from the site of Washington, D.C. Since this example, if offered alone, might serve merely to confirm the commonly held view that description by metes and bounds was pe-

---

[1] Ibid.

[2] Gibson, *Theory and Practice of Surveying*, p. 352.

[3] Flint, op. cit., p. 77. Flint was under no illusion about Polaris' position relative to the North Celestial Pole. He described methods for the determination of true north by reference to Polaris at elongation and culmination. (Ibid., pp. 77-78.)

[4] Charles W. Stetson, *Four Mile Run Land Grants* (Washington: Mimeoform Press, 1935), p. 90.

culiar to areas where the practice of indiscriminate locations prevailed,[1] the following extract from a deed conveying property within a New England township should also be placed on record. Dated 1735, it reads:

> . . . beginning at the Foot of the Gulley below his House and running three hundred and twenty Pole North Five Degrees West to Red Oak marked AB. then running Eighty three Pole East Five Degrees North to a Spruce Tree mark^d AB then running South three hundred and twenty Pole to a Pitch Pine mark^d then running Fifty three Poles West and by South which makes up the one hundred & thirty four Acres. . . .[2]

New England lots were also, on occasion, described by reference to adjoining property, and to distinguishing internal characteristics.[3] Finally, in the decades immediately preceding the opening of the national public domain, the laying out of New England townships with lots bounded by right lines permitted the adoption of numbering systems whose simplicity was comparable to that proposed in the report of Jefferson's committee.[4]

The provision for numbering lots, in the national ordinance, read as follows (see Fig. 7B):

> . . . the said lots . . . in every Hundred shall be designated by the numbers in their order from 1. to 100. beginning at the Northwestern lot of the Hundred and applying the numbers from 1. to 10. to the lots of the first row from West to East successively, those from 11. to 20. to the lots of the second row from West to East and so on.[5]

---

[1] Description by metes and bounds is associated with surveying under the system of indiscriminate locations, for example, in Charles O. Paullin, Atlas of the Historical Geography of the United States, ed. John K. Wright, Carnegie Institution of Washington Publication No. 401 (Washington: Carnegie Institution of Washington, 1932), p. 25.

[2] From deed to land in Town of Scarborough, York County, Massachusetts (now Maine), in H. W. Richardson et al. (eds.), York Deeds (18 vols.; Bethel, Maine, 1903-1910), XVII, 260.

[3] See ibid., passim; William B. Trask et al. (eds.), Suffolk Deeds (14 vols.; Boston, 1880-1906), Vol. X, passim; Report of the Commissioners Appointed to Complete the Examination and Determination of All Questions of Title to Land . . . on the Isle of Martha's Vineyard (Boston, 1871), Appendix, passim; and Charles M. Andrews, The River Towns of Connecticut, Johns Hopkins University Studies in Historical and Political Science, Seventh Series, Nos. 7, 8, 9 (Baltimore: Johns Hopkins University, 1889), p. 44.

[4] See plats of townships granted in present day Vermont by the Mason Proprietary subsequent to 1846, in New Hampshire State Papers, XXVII, 180, 200, 266, 318, 330, 450, 498.

[5] Boyd, Papers of Thomas Jefferson, VII, 141.

Interestingly, vestiges of older means of identifying lots were retained in the law. In part, this was due to the necessity of identifying claims in advance of survey. Lots were to be identified, should claims be entered in advance, "by a designation of some point, either natural or artificial, within the said lot . . . so singular and certain as may be adapted to no other. . . ."[1] In suggesting that a lot be identified in a similar fashion even in the final deed, however, the proposed ordinance probably submitted to the force of habit.[2] With respect to hundreds, the committee report generally adhered to the idea of "particular marks, natural or artificial," though it also anticipated "stating the order or position of the Hundred relatively to the boundaries of the state."[3] The sufficiency of numbers alone was recognized in the land ordinance of the following year.

Review

With respect to surveying, the proposed land ordinance of 1784 largely bears out Treat's contention that, in the legislation of the period, "few things were done de novo."[4] Certainly the duties of registers and surveyors, and the use of marked trees and the surveyor's chain and plat represented carry-overs from established practice. But this view tends to obscure the importance of such reforms in the law as insistence on the correction of magnetic bearings, and the curtailment of description by metes and bounds, not to mention the Jefferson-Williamson grid. As to precedents, a leading thesis of the present study is that the Southern land system comprised the basis of the proposed land ordinance of 1784, and that New England precedents found their expression in the following year. "Some credit surely belongs," Treat wrote, "to the men who, in 1785, perfected the rough plan and made it into law."[5] To give them due credit will be the aim of the following chapter.

---

[1] Ibid.

[2] Recommendation of identification by description was combined with that of identification by simple numbering. (Ibid.)

[3] Ibid.

[4] Treat, "Origin of the National Land System," p. 234.

[5] Treat, National Land System, p. 182.

CHAPTER IV

THE LAND ORDINANCE OF 1785

"We wait with impatience," wrote David Howell in early February, 1785, "for the result of the negociations opened with the Indians." "It is expected that Congress, before they rise," he added, "will be enabled to open their land office."[1] A few days later, official news of the Treaty of Fort McIntosh reached Congress.[2] On March 4, the report of Jefferson's committee on public lands was once more brought before Congress, and then submitted to a new committee, headed by William Grayson of Virginia.[3] On April 4, the committee's report, "drawn principally by Colo. Grayson,"[4] was read a first time, and for over a month thereafter the attention of Congress was intermittently taken up with its consideration.[5] Howell, during this period of debate, found the land ordinance "the most complicated and embarrassing subject before Congress since peace has taken place."[6] As Grayson explained

---

[1] Howell to Governor of Rhode Island, New York, February 9, 1785, in Staples, op. cit., p. 524. Congress was now meeting in New York City, where it sat from January 11, 1785, to November 4, 1785.

[2] "The commissioners for treating with the western Indians did yesterday present to Congress the Treaty that they have made with the Wyandots, Delawares, etc." Richard Henry Lee to George Washington, New York, February 14, 1785, in Burnett, Letters of Members, VIII, 36.

[3] The report of Jefferson's committee was "read a first time," March 4. Jrnls. Cont. Cong., XXVIII, 114. A second reading and the appointment of the new committee took place on March 16. Ibid., p. 165.

[4] Monroe to Jefferson, New York, April 12, 1785, in Hamilton, Writings of James Monroe, I, 71.

[5] The committee reported April 12. Jrnls. Cont. Cong., XXVIII, 251-256. After a first reading April 14 (ibid., p. 264), the report seems to have been recommitted, but on April 22 Congress "proceeded in the consideration of the Ordinance" (ibid., p. 290), and continued to debate the measure through May 6 (ibid., pp. 290-343, passim). A "third reading" of the Ordinance was recorded for both May 6 and May 19 (ibid., pp. 342-370).

[6] Howell to the Governor of Rhode Island, New York, April 29, 1785, in Staples, op. cit., p. 528.

toward the end of the debates, there were never "above ten States on the floor and nine of these were necessary to concur in one sentiment" respecting each provision in the law, "lest they should refuse to vote for the Ordinance on it's passage."[1] At last, May 20, 1785, the Land Ordinance was passed by the unanimous vote of the states represented.[2]

Behind the determination of Congress to reach agreement and enact a land ordinance lay the hope of removing a burdensome public debt. It might be observed, in this connection, that the question of land disposal was no longer coupled, in 1785, with the problem of government in the West. With the latter concern temporarily disposed of, by the Ordinance of 1784, the new land ordinance found itself sharing congressional attention with the almost desperate problem of federal finance. "Land Office and Requisition now occupy us," wrote Rufus King, in the middle of April.[3] By requisition King meant a levy by Congress upon the states for contributions in support of the federal government. The annual requisition was presented in Congress two weeks before Grayson's committee reported a land ordinance,[4] and its consideration, complicated more than usual by unsettled accounts and interstate jealousies, continued after the Land Ordinance was passed.[5] Of the amount called for in the requisition of 1785, over two-thirds was to be applied to the payment of interest on the domestic debt.[6] Richard Henry Lee expressed a general view when he

---

[1] Grayson to Washington, New York, May [8], 1785, in Burnett, Letters of Members, VIII, 118. The assent of nine states was required by Article IX of the Articles of Confederation for the passage of certain categories of legislation.

[2] No vote is recorded in Jrnls. Cont. Cong. One group of delegates reported passage of the Ordinance "by the unanimous voice of all the States present." New Hampshire delegates to President of New Hampshire, New York, May 29, 1785, Burnett, Letters of Members, VIII, 124. Two of the thirteen member states, Delaware and North Carolina, were not represented. (Jrnls. Cont. Cong., XXVIII, 365-375.) For text of the Ordinance, see ibid., pp. 375-381, and Carter, Territorial Papers, II, 12-18.

[3] Rufus King to Elbridge Gerry [New York], April 18, 1785, in Burnett, Letters of Members, VIII, 98.

[4] Report of committee on requisition, March 31, 1785, Jrnls. Cont. Cong., XXVIII, 214-220.

[5] For a brief account of the progress of the requisition through Congress, see Burnett, The Continental Congress, pp. 619-622.

[6] Jrnls. Cont. Cong., XXVIII, 215-216.

referred to the federal lands as "this fine fund for extinguishing the public debt."[1] The basic importance of this view was affirmed by Grayson, who wrote, "If the importunities of the public creditors, and the reluctance to pay them by taxation either direct or implied had not been so great I am satisfied no land Ordinance could have been procured."[2]

The present chapter will be entirely concerned with the Land Ordinance of 1785, but, as in preceding chapters, the approach will be specialized. Attention will center upon an interpretation of the surveying provisions in the law, and questions of land disposal will be treated in an incidental fashion. As a preliminary to interpretation, some notice should be given to the committee of 1785 which revised the report of Jefferson's committee and developed the greater part of the final Ordinance. This was a "grand committee," composed of one delegate from each of

---

[1] Lee to Washington, New York, April 18, 1785, in Burnett, Letters of Members, VIII, 98. Similar views were expressed by other delegates, among them, Hardy, Johnson, Howell and Monroe, for whose opinions see ibid., pp. 85, 101, 106 and 117, respectively. The committee on the requisition for 1785 added to its report, "As a motive for cheerful payment of this requisition . . . the Committee are of opinion that the States be informed that Congress are about soon to open a Land Office to dispose of their Western Territory, and that the proceeds thereof will be applied as a sinking fund to extinguish the principal of the domestic debts." (Jrnls. Cont. Cong., XXVIII, 220.) This was not a new proposition. The land ordinance reported by Jefferson's committee in 1784 required that the revenue arising from the sale of lands "shall be applied to the sinking such part of the principal of the national debt as Congress shall from time to time direct." (Ibid., XXVII, 451.) Earlier Jefferson had been opposed to the disposal of lands for gain. The revenue motive is contrasted with colonial tradition in Henry Tatter, "State and Federal Land Policy during the Confederation Period," Agricultural History, IX (October, 1935), 176-186.

[2] Grayson to Washington, New York, May [8], 1785. Regarding resistance in Congress to passage of the Ordinance, Grayson wrote, "Several of the States are averse to new votes from that part of the Continent and . . . some of them are now disposing of their own vacant lands, and of course wish to have their particular debts and their own countries settled in the first instance before there is any interference from any other quarter." (Ibid.) Operating against this resistance, in addition to desire for revenue, were (1) demands for military bounty lands, (2) the need for frontier defense, and (3) the pressure of immigration to the West. These and other inducements to the formation of a national land policy are conveniently enumerated in Benjamin Horace Hibbard, A History of the Public Land Policies (New York: The Macmillan Company, 1924), pp. 32-35. A case for the basic importance of the demands for military bounty lands is made in Rudolph Freund, "Military Bounty Lands and the Origins of the Public Domain," Agricultural History, XX (January, 1946), 8-18.

the states represented in Congress. It included only Williamson and Howell from the committee of the preceding year.[1] The two members who stood forth in the new committee were Grayson and King.[2] Grayson's role was clearly that of an expediter. He set forth, in a letter justifying the committee's report, not his own views, but "the reasons which those who are advocates for the measure offer in it's support."[3] To him must go the credit for pushing a bill which included what were to him "exceptionable parts."[4] Rufus King of Massachusetts, on the other hand, figured as a positive agent in the alteration of the Jefferson report. In a letter, he assured Gerry, who had left Congress, "When I tell you the History of this ordinance you shall acknowledge that I have some merit in the business."[5]

The Land Ordinance of 1785 was satisfactorily summarized by James Monroe, shortly before its passage, in the following words:

> [The territory] is to be survey'd in townships containing abt. 26,000 acres each, each township mark'd on the plat in-

---

[1] The committee members were: Pierse Long (New Hampshire), Rufus King (Massachusetts), David Howell (Rhode Island), William Samuel Johnson (Connecticut), Robert R. Livingston (New York), Archibald Stewart (New Jersey), Joseph Gardner (Pennsylvania), John Henry (Maryland), William Grayson (Virginia), Hugh Williamson (North Carolina), John Bull (South Carolina), and William Houston (Georgia). (Jrnls. Cont. Cong., XXVIII, 251.)

[2] William Grayson, Revolutionary soldier, lawyer and former member of the Virginia House of Delegates, first took his seat in the Continental Congress in March, 1785. He was later United States Senator from Virginia. For biographical sketch, see Dictionary of American Biography, Vol. VII. Rufus King served in the Continental Congress from 1784 to 1786. Later, after playing an important part in the Constitutional Convention, he served twice as United States Senator from New York, and twice as Ambassador to Great Britain. See ibid., Vol. X, and Charles R. King (ed.), The Life and Correspondence of Rufus King, Comprising His Letters, Private and Official, His Public Documents and Speeches (6 vols.; New York: G. P. Putnam's Sons, 1894-1900).

[3] Grayson to Washington, New York, April 15, 1785, in Burnett, Letters of Members, VIII, 95.

[4] Ibid., p. 96. Grayson took charge of the bill after its emergence from committee. (Grayson to Timothy Pickering, New York, April 27, 1785, ibid., p. 106.) When the ordinance was well advanced toward passage, Grayson wrote, "I am sorry to observe that throughout this measure, there has been a necessity for sacrificing one's own opinion to that of other people for the purpose of getting forward." (Grayson to Washington, New York, May [8], 1785, ibid., p. 118.)

[5] King to Gerry, New York, May 8, 1785, in King, Life of Rufus King, I, 94.

to lots of one mile square, and 1/2 the country sold only in townships and the other in lots. 13 surveyors are to be appointed for the purpose to act under the controul of the Geographer, beginning with the first range of townships upon the Ohio and running North to the lakes, from [a point due north of] the termination of the line which forms the southern boundary of the State of Pena., and so on westward with each range. As soon as . . . [seven] ranges shall be survey'd, the return will be made to the Bd. of Treasury, who are instructed to draw for them in the name of each State in the proportion of the requisition on each, and transmit its portion to the loan officer in each, for sale at public provided it is, nor any part, sold for less than one doll'r specie or certificates the Acre.[1]

The surveying content of the Ordinance, toward a fuller appreciation of which discussion in earlier chapters has been directed, will now be considered in detail, under six headings.

The Principle of Prior Survey

The first alteration of the Jefferson report to attract the attention of Monroe, when he read the ordinance as prepared by Grayson's committee, was the adoption of the principle of prior survey. Writing to Jefferson in Paris, he said, "It deviates I believe essentially from the one . . . [of last year]. . . . [T]he object of this is to have the lands survey'd previous to the sale, and after the survey to have the lots . . . sold . . . at public vendue [auction]."[2] The Jefferson-Williamson grid had been taken from its original context and deprived of its original function as a control over the southern land system. The southern system of land disposal had been abandoned. "The present plan," Grayson explained to Washington at about the same time as Monroe addressed Jefferson, "excludes all the formalities of warrants entries locations returns and caveats, as the first and last process is a deed."[3]

Grayson, expecting that prior survey and sale at public auction might appear to Washington "at first view eccentric and objectionable," offered the following, in defense:

> They say that this cannot be avoided with't affording an undue advantage to those whose contiguity to the territory has given them an opportunity of investigating the quality of

---

[1] Monroe to Madison, New York, May [8], 1785, in Hamilton, *Writings of James Monroe*, I, 77.

[2] Monroe to Jefferson, New York, April 12, 1785, *ibid.*, p. 71.

[3] Grayson to Washington, New York, April 15, 1785, in Burnett, *Letters of Members*, VIII, 96.

the land; that there certainly must be a difference in the value of the lands in different parts of the country, and that this difference cannot be ascertained with't an actual survey in the first instance and a sale by competition in the next.[1]

Grayson was quoting "the advocates for the measure," as was his practice generally in this letter of explanation. Prior survey and public auction comprised a mode of land disposal earlier authorized in Connecticut, New York and Massachusetts.[2] Timothy Pickering of Massachusetts, not a member of Congress,[3] had encouraged the adoption of this procedure into federal law. In an advisory letter of which Rufus King was in possession when Grayson's committee convened, Pickering said he looked forward to the surveying of townships first, "and then selling these townships at public auction."[4] In later sending a copy of the Gray-

---

[1] Ibid., p. 95.

[2] See act of Connecticut's General Assembly, 1737, governing sale and settlement of townships in western Connecticut, in C. J. Hoadley (ed.), Public Records of the Colony of Connecticut (15 vols.; Hartford: The Press of the Case, Lockwood and Brainard Co., 1850-1890), VIII, 134-137. In New York, an act of May 12, 1784, governing the disposal of lands confiscated during the Revolutionary War, provided for sale "at public vendue, to the highest bidder or bidders" or at private sale, whichever might be deemed most beneficial to the state. (Laws of the State of New-York . . . from the First to the Twentieth Session, Inclusive [3 vols.; New York: Printed by Thomas Greenleaf, 1798], I, 128.) Similarly, Massachusetts, by a law passed in July, 1784, provided for the sale of confiscated lands "at public auction." (Acts and Laws of the Commonwealth of Massachusetts [13 vols.; Boston: Printed by Wright and Potter Printing Co., 1890-1898], I, 234.) In June, 1785, after passage of the national land ordinance, lands in western Massachusetts were ordered sold "either at public or private sale." (Ibid., p. 660.)

[3] Pickering, whose remarks on the convergence of meridians have been quoted earlier, was at this time engaged in the mercantile business in Philadelphia, after having served as Quartermaster General from 1780 to the close of the Revolutionary War. After the adoption of the Constitution, Pickering held three cabinet posts under President Washington, and later served as United States Senator from Massachusetts, 1803-1811. See Dictionary of American Biography, Vol. XIV, and Octavius Pickering, op. cit.

[4] Pickering to Elbridge Gerry, Philadelphia, March 1, 1785, ibid., I, 504-505. Gerry handed this letter to King, before leaving Congress. (Ibid., p. 506.) Earlier, in 1781, Pelatiah Webster, Connecticut-born political economist, had outlined a national land system which included survey before sale, as well as several other features later embodied in the Land Ordinance of 1785: see Pelatiah Webster, Political Essays on the Nature and Operation of Money, Public Finances, and Other Subjects (Philadelphia, 1791), pp. 481-500.

son report to Pickering, King remarked, "You will find thereby that your ideas have had weight with the committee."[1]

The term prior survey may suggest to the reader the policy of surveying state boundaries before proceeding with subdivision and settlement. This policy, laid down in the Ordinance of 1784, and further developed in the land ordinance intended to accompany it, was ignored in the Land Ordinance of 1785. This is not to say, however, that Jefferson's scheme for bounding western states was without significance to the history of public land surveying. Of this significance, already discussed, a reminder will be included under the heading which appears next, below.

The Rectilinear Grid Retained

Preserved from the report of Jefferson's committee was the provision that western lands be divided "by lines running due north and south and others crossing these at right angles."[2] The Jefferson-Williamson plan for rectilinear surveying was stripped of its hundreds, its geographical mile, and its decimal divisions. Now, townships were to be formed, each six statute miles square, but the essential grid remained, as the most important legacy from the committee report of 1784. In justification of the grid, two new and strong reasons were advanced: that it would be "attended by the least possible expence, there being only two sides of the square to run in almost all cases," and that there would be "exemption from controversy on account of bounds to the latest ages."[3]

Inherent in the grid, of course, was the conflict between rectangularity and the convergence of meridians. This was somewhat alleviated by an allowance in the Ordinance for lines to en-

---

[1] King to Pickering, New York, April 15, 1785, in King, Life of Rufus King, I, 46. King added, "I shall hold myself particularly obliged by you for these communications on the subject." King referred not only to the letter of March 1, addressed to Gerry, but to two letters of March 8, addressed to himself. These appear in part, ibid., pp. 43-46. Pickering's most important recommendation, in these letters, was that of legislation for the exclusion of slavery from the West. A motion with this objective was made by King, March 16. A similarly expressed prohibition was later embodied in the Northwest Ordinance of 1787.

[2] Carter, Territorial Papers, II, 13; Jrnls. Cont. Cong., XVIII, 375.

[3] Grayson to Washington, New York, April 15, 1785, in Burnett, Letters of Members, VIII, 95-96.

close six-mile squares while maintaining their cardinal orientation, "as near as may be."[1] The reader will recall, from the discussion of this conflict as it first appeared in the Jefferson-Williamson plan, that the eventual solution of the difficulty lay in the employment of prime meridians and base lines (Fig. 8), which simply recaptured the original utility of Jefferson's abandoned state boundaries (Fig. 4).

The rectilinear scheme of survey had its opponents. From North Carolina came Richard Spaight's voice, grumbling against "this formal and hitherto unheard of plan."[2] Said George Washington, "the lands are of so versitile a nature, that to the end of time they will not, by those who are acquainted therewith, be purchased either in Townships or by square miles."[3] These were southern protests. The most important threat came from New England. The only two efforts to disrupt the rectilinear grid through later amendment of the Land Ordinance originated with delegates from Massachusetts. The first of these was expressed in a resolution moved by Nathan Dane, May 3, 1786:

> That in dividing the said territory into townships due regard be had to the natural boundaries of Townships of those particular cases wherein a rigid adherence to lines run East and West, North and South, as boundaries would manifestly prejudice the sales and future condition of said townships.[4]

This resolution, apparently not voted upon, serves as a reminder that New England tradition, despite a general tendency toward square townships, did not countenance the continuous use of completely arbitrary lines, and that the strict pattern of national

---

[1] Carter, Territorial Papers, II, 13; Jrnls. Cont. Cong. For earlier discussion, see section headed "Rectangles and Meridians," chap. ii, above.

[2] Richard Dobbs Spaight to the Governor of North Carolina, New York, June 5, 1785, in Burnett, Letters of Members, VIII, 135. Spaight expected surveying to progress so slowly that the lands would be possessed "by persons, who have already and are daily crossing the Ohio, in great numbers, so as to put the United States to more expence to dispossess them, than the soil will afterwards sell for." Ibid., pp. 135-136.

[3] Washington to Grayson, Mount Vernon, August 22, 1785, in Fitzpatrick, Writings of George Washington, XXVIII, 234. The next year, Washington wrote, "I had, and still have my doubts of the utility of the plan, but pray devoutly that they may never be realized, as I am desirous of seeing it a productive branch of the Revenue." Washington to Grayson, Mount Vernon, July 26, 1786, ibid., p. 486.

[4] Jrnls. Cont. Cong., XXX, 231.

surveying found its origin elsewhere.[1] The effect of license to employ natural boundaries was illustrated by the contemporaneous work of Rufus Putnam. Putnam laid out seven townships between the Schoodic and Cobscook rivers in the present state of Maine, in 1784.[2] The townships, which were to "contain 6 miles square,"[3] were surveyed as shown upon the accompanying map (Fig. 9A), which serves to suggest the degree of distortion which would, occasionally, have resulted, if Congress had adopted Dane's resolution.

The second move to alter the grid, through amendment of the Land Ordinance, was made by Rufus King, May 12, 1786. King asked the repeal of a provision, carried over from the Jefferson report, that all lines be run "by the true meridian."[4] This motion, again, simply aimed at rendering the federal surveys conformable to New England practice. One from a series of New England townships, relatively new at the time of King's motion, illustrates the inaccuracy of lines run by the magnetic needle without reference to true north. This series, in the southern part of present-day Vermont, has been accorded attention in earlier studies for the resemblance of its township boundaries to the pattern of the national land system.[5] With the effect of

---

[1] The failure of New England to furnish a precedent for the strict pattern of the national surveys is made evident in Amelia Ford, op. cit., pp. 28-42. The nearest approach to a precedent is discussed in the next paragraph, below. Following the passage of the national land act, one block of townships was laid out in Maine, with striking regularity, apparently where it was thought that a regular grid would not "prejudice the sales and future condition" of the townships. This block, newly surveyed and isolated, is a conspicuous feature of Osgood Carleton's "Map of the District of Maine," in James Sullivan, The History of the District of Maine (Boston, 1795), frontispiece. The map is reproduced in Paullin, Atlas of the Historical Geography of the United States, Plate 44A. It should be noted that the grid shows no regard for orientation to the cardinal points of the compass.

[2] Report of the Committee for the Sale of Eastern Lands: Containing Their Accounts from the 28th of October, 1783 to the 16th of June, 1795 [Boston, 1795], p. 4.

[3] Resolves of the General Court of the Commonwealth of Massachusetts, Respecting the Sale of Eastern Lands; with the Reports of the Committees Appointed To Sell Said Lands; from March 1, 1781 to [June 22, 1803] (Boston, 1803), p. 30.

[4] Jrnls. Cont. Cong., XXX, 262.

[5] Governor Benning Wentworth of New Hampshire issued over one hundred charters for townships west of the Connecticut River between 1749 and 1764. In calling for six-mile townships these charters did not differ from many others, cited in W. A. Truesdell, "The Rectangular System of Surveying," Jrnls. Assn. Engin.

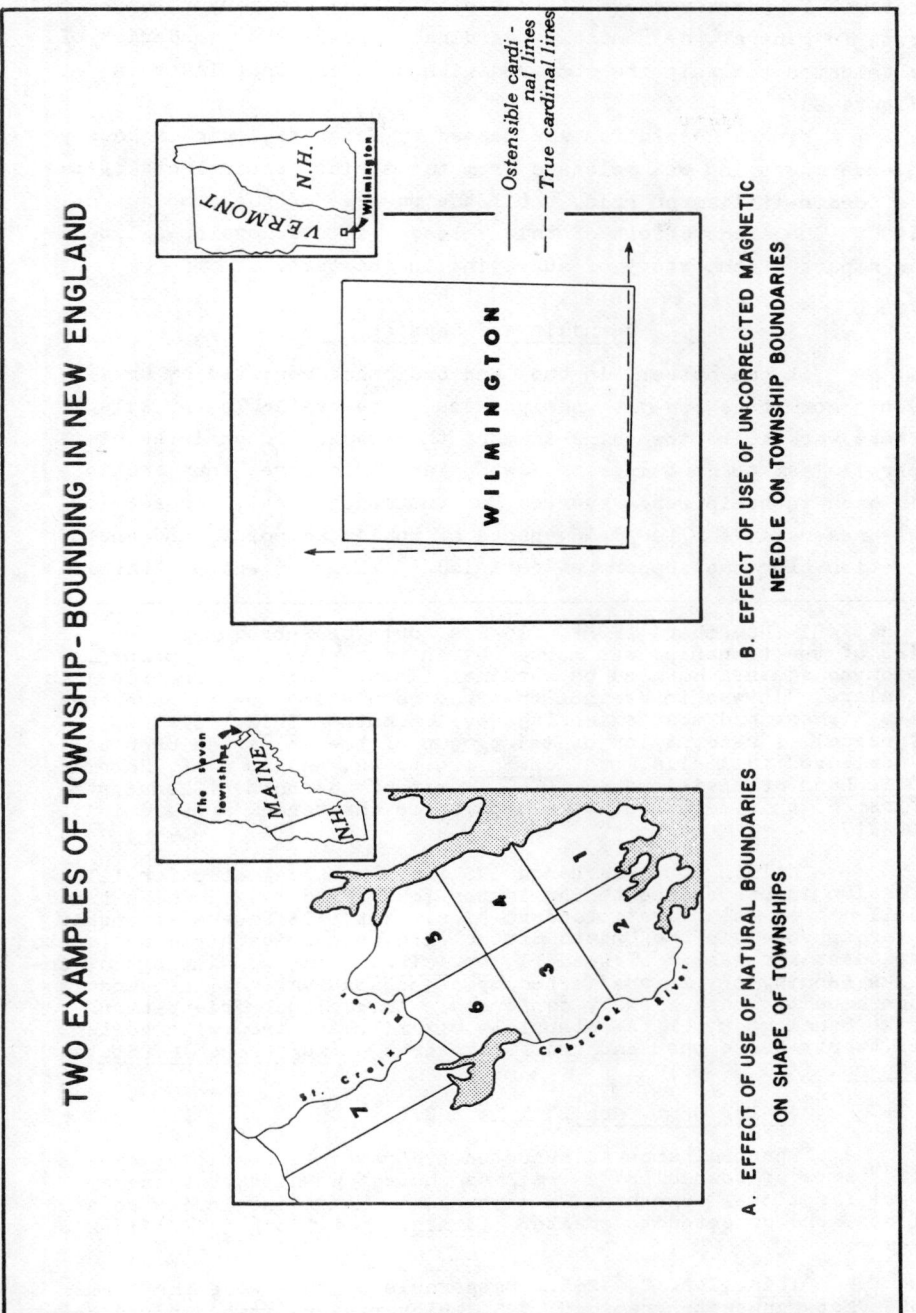

Fig. 9

natural features reduced, if not eliminated, the boundaries were run, in general, as nominally cardinal lines. The boundaries of a selected township are compared with true cardinal lines in Figure 9B.[1]

King's resolution was passed by Congress,[2] and in consequence surveying was released from the strict requirements of the Jefferson-Williamson grid, until the passage of the Land Act of 1796. The exact effect of this release will be developed later, as a part of the story of surveying in the West.

Township and Section

At the outset, in the land ordinance reported by Grayson's committee, only townships were to be available for sale.[3] These were to be townships in more than name. In unmistakable parallelism to contemporary New England procedure, four sections in each township were reserved for future disposal, one section was reserved "for the maintenance of public schools," and another section "for the support of religion."[4] Grayson wrote, "The idea

---

Soc., XLI (November, 1908), 216-219, but the subsequent surveying of the townships was marked by an unusually close approximation to squares bounded by cardinal lines. This led Truesdell to declare, "It was in Vermont that the regulation township reached its highest and most extensive development." Ibid., p. 219. Truesdell's recognition of this group of townships has been acknowledged in Amelia Ford, op. cit., p. 42, and in C. E. Sherman, Ohio Land Subdivisions, p. 216. A map of the Town of Bennington, first of this series of townships to be chartered, appears ibid., p. 217.

[1] Wilmington (chartered 1751) has been selected for illustration in preference to Bennington (chartered 1749), which Truesdell called "the first standard [i.e., six mile square] township ever surveyed in the United States with boundaries north and south and east and west." Truesdell, op. cit., p. 219. Wilmington, unlike Bennington, is unaffected by an adjacent state line, and is consequently more closely conformable to cardinal orientation. Like Bennington, it is ostensibly six miles square, with contents of twenty-three thousand and forty acres. (New Hampshire State Papers, XXVI, 736.)

[2] Jrnls. Cont. Cong., XXX, 262.

[3] The ordinance as reported by Grayson's committee speaks of "parts of townships" as well as whole townships, but these were fractional townships which the Ohio River and Indian cession lines were expected to create. (Jrnls. Cont. Cong., XXVIII, 253, 254.)

[4] Ibid., pp. 254-255. Comparable to this were the terms of a Massachusetts grant of 1785, which reserved two hundred acres in each township for future disposition, four hundred acres for

of a township, with the temptation of a support for religion and education, holds forth an inducement for neighborhoods of the same religious sentiments to confederate for the purpose of purchasing and settling together."[1] Consistent with a suggestion of Pickering's, the report proposed that a part of "all gold, silver, lead, Copper and Coal mines, and all salt licks and salt springs" be reserved.[2] In the Grayson report, with these provisions, and the requirement of prior survey and sale at public auction, the New England influence was at high tide.

In subsequent debates, the New England delegates, "amazingly attached to their own customs,"[3] held to the proposals of the Grayson report. Although King claimed that the southern delegates were "for indiscriminate Locations etc.,"[4] the Journals of Congress suggest rather that they had accepted the defeat of that system, and were fighting only for that purchaser's prerogative which the Jefferson-Williamson plan had allowed. They wanted the sale of small quantities within the framework of the grid.[5] From the New England point of view, this boded the "same consequence" as indiscriminate locations: "a tendency to destroy all those inducements to emigration which are derived from friendships, religion and relative connections."[6] Sale by whole townships, on the other hand, from the southern point of view, meant the unwelcome necessity of purchasing land "rough as it runs,"[7] that is, good

---

church and minister, and two hundred and eighty acres for a grammar school. (Resolves of the General Court of Massachusetts [1803], pp. 27-28.)

[1] Grayson to Washington, April 15, 1785, in Burnett, Letters of Members, VIII, 95.

[2] Jrnls. Cont. Cong., XXVIII, 254. Pickering's suggestion appears in Pickering to King [Philadelphia], March 8, 1785, in King, Life of Rufus King, I, 44. A similar suggestion was made by George Washington. (Washington to the President of Congress, Mount Vernon, December 14, 1784 in Fitzpatrick, Writings of George Washington, XXVIII, 11.)

[3] Grayson to Madison, New York, May 1, 1785, in Burnett, Letters of Members, VIII, 109-110.

[4] King to Gerry, New York, April 26, 1785, in King, Life of Rufus King, I, 91.

[5] There are no recorded motions in the Journals calling for other than gridded subdivision.

[6] Grayson to Washington, New York, April 15, 1785, in Burnett, Letters of Members, VIII, 95.

[7] Grayson to Madison, New York, May 1, 1785, ibid., p. 95.

and bad together, and of course left no opportunity for direct purchase by the individual settler. Although quantities within a township both less than and greater than a square mile were suggested, southern opinion generally favored the square mile section.[1] Interestingly, it was Howell, seconded by Williamson, who first moved the opening of the entire territory to sale by sections, as in the Jefferson report.[2] These were no longer square geographical miles, of course, but 640 acre units, for which there was colonial precedent.[3]

In the final Ordinance, an obvious compromise was adopted. Townships were ordered to be sold alternately entire and by section.[4] Sections (called "lots" in the final Ordinance)[5] were not to be surveyed. Surveyors were directed to mark corners "at the interval of every mile" along the boundaries of the townships,

---

[1] "The Southern people . . . were for selling the whole territory in lots of a mile square." Grayson to Madison, New York, May 28, 1785, in Burnett, Letters of Members, VIII, 130.

[2] There was, however, this limitation: sections were to be sold in the order of their number on the plat, and a second township could not be sold by sections until the whole of the first had been taken up. (Jrnls. Cont. Cong., XXVIII, 290-291.) Howell later offered a modified version of this motion. (Ibid., p. 336.) Williamson later moved the exposure to sale by section of such townships as had not been sold entire within a certain number of months after the opening of sales. (Ibid., p. 371.)

[3] These precedents are reviewed in Amelia Ford, op. cit., pp. 43-53. For further evidence of the prevalence of this unit and conveniently related divisions in North Carolina, see William K. Boyd (ed.), Some Eighteenth Century Tracts Concerning North Carolina (Raleigh, North Carolina: Edwards and Broughton Co., 1927), p. 443, and John Love, The Whole Art of Surveying and Measuring of Land Made Easie (3d ed.; London, 1716), pp. 2, 132-136. Frederick Jackson Turner, perhaps on the basis of Miss Ford's research, spoke of the 640-acre unit in federal surveying as though it were an inherited characteristic, which had passed from North Carolina law through Kentucky's frontier stations into the national land system. See Frederick Jackson Turner, "The Old West," Proceedings of the State Historical Society of Wisconsin, 1908 (Madison: State Historical Society of Wisconsin, 1909), pp. 231-232.

[4] Carter, Territorial Papers, II, 15. Grayson said that he offered such a compromise early in the debates, "under the impression that it would accomodate both the Eastern and Southern States," but without success. Grayson to Pickering, New York, April 27, 1785, in Burnett, Letters of Members, VIII, 106. The motion does not appear in Jrnls. Cont. Cong.

[5] The disappearance and subsequent reappearance of the term section is noted under the heading, "The Square-Mile Section," chap. ii, above.

but internal lines were to be shown on paper only.[1] This was a regression from the square-mile lot as proposed in the Jefferson report, and from the original compromise motion as well, wherein sections were to be surveyed.[2] Presumably, it was felt that "the expence and delay would be too great," this having been the reason given for not authorizing such subdivision in the Grayson report.[3] The omission of surveyed subdivision was rectified in the Land Act of 1796.[4] It was not until 1800, when frontiersmen had gained a voice in Congress, that quantities of less than a square mile were ordered to be surveyed and made available for sale.[5]

In arriving at a final compromise, an attempt made by King to extend federal surveying into the Virginia Military Reserve was rebuffed,[6] certain changes were made in the designation of general and mineral reserves,[7] and the provision for reserving

---

[1] Carter, *Territorial Papers*, II, 13-14; *Jrnls. Cont. Cong.*, XXVIII, 376.

[2] Grayson expected to "double the quantity of surveying," in offering this compromise. Grayson to Pickering, New York, April 27, 1785, in Burnett, *Letters of Members*, VIII, 106.

[3] Grayson to Washington, New York, April 15, 1785, *ibid.*, p. 96.

[4] By this act, to be discussed in a later chapter, townships were ordered subdivided by lines to be run at two mile intervals. Carter, *Territorial Papers*, II, 553, 554.

[5] By the Land Act of 1800, alternate townships west of the Muskingum River were ordered subdivided into 320-acre lots. (Carter, *Territorial Papers*, III, 89.) Even in 1785, attempts had been made to authorize subdivisions of this size, in fractional townships. (*Jrnls. Cont. Cong.*, XXVIII, 253, 343.)

[6] King apparently aimed at controlling Virginia's claims within the Reserve by placing survey and disposal wholly in the hands of the federal government. Since he recognized the right of Virginia's troops to "good lands," it is not clear whether or not he contemplated surveys that would break the pattern of the grid in enclosing such lands. (*Jrnls. Cont. Cong.*, XXVIII, 309-310.) Grayson, however, confirms that some members of Congress favored extension of the grid into the Reserve. (Grayson to Madison, New York, May 1, 1785, in Burnett, *Letters of Members*, VIII, 110.) The day following King's motion, April 28, a counter-motion by the Virginia delegates was approved, which by implication reserved the right of survey to individual claimants. (*Ibid.*, pp. 316-317.) For this amendment in final Ordinance, see *ibid.*, p. 381, and Carter, *Territorial Papers*, II, 18.

[7] In the final Ordinance, the position of lots to be generally "reserved for the United States" was simply shifted from each of the four corners of a township to the center of each

land in support of religion was eliminated.¹ Established by the Land Ordinance was the proposition that "there shall be reserved the lot N 16 of every township for the maintenance of public schools."² Finally, the size of townships, seven miles square in the Grayson report, was reduced to six miles square.³

Surveying and Numbering

Two paragraphs in the Land Ordinance of 1785 governed general surveying procedure. They were drawn from the Jefferson report. On the orientation of lines, this paragraph was transferred, with negligible alteration:

> The geographer and surveyors shall pay the utmost attention to the variation of the magnetic needle, and shall run and note all lines by the true meridian, certifying with every plat what was the variation at the times of running the lines thereon noted.⁴

---

quarter of a township. In addition, "one third part of all gold, silver, lead and copper mines" was reserved from immediate sale. (Carter, Territorial Papers, II, 15, and Jrnls. Cont. Cong., XXVIII, 378.) Interestingly, salt springs and salt licks were removed from the reserved list, whereas in the Grayson report these, together with "a square of one hundred acres of land, of which the said salt lick or salt spring shall be the centre" were reserved. (Ibid., p. 254.)

[1] For the vote on this part of the law, which was not divided along regional lines, see Jrnls. Cont. Cong., XXVIII, 295.

[2] Carter, Territorial Papers, II, 15. In the Grayson report, the center section of each township was to be reserved for this purpose. With the reduction in the size of townships from seven to six miles square (see following note), the center section disappeared.

[3] In the report of Grayson's committee, provision was made for seven-mile townships and then cancelled. (Jrnls. Cont. Cong., XXVIII, 252.) A revised version of the ordinance, April 26, repeated the provision (Ibid., p. 298.) That delegates from New York were responsible for the seven-mile dimension is suggested by the fact that a New York law, passed in 1781, provided for the grouping of military bounty lands into townships seven miles square. (Laws of the State of New York [1798], I, 41.) Further, when the amendment to reduce the size of townships was voted upon, New York was the only state to oppose the change. (Jrnls. Cont. Cong., XXVIII, 327.) Curiously, delegates from New England, where the six-mile township was common, were said to have adhered strongly to the idea of the seven-mile township, once it had been proposed. Grayson to Madison, New York, May 28, 1785, in Burnett, Letters of Members, VIII, 129-130, and Monroe to Madison, New York, May [8], 1785, in Hamilton, Writings of James Monroe, I, 77.) The motion to amend was made by Grayson and seconded by Monroe.

[4] Carter, Territorial Papers, II, 14. See also Jrnls. Cont. Cong., XXVIII, 376-377.

This directive, which has received full consideration earlier in this study, as a part of the Jefferson report,[1] would have forced the conflict between rectangularity and convergency, if allowed to stand. It was repealed, as has been said, in May, 1786.

The second paragraph on the conduct of surveying read as follows:

> The lines shall be measured with a chain; shall be plainly marked by chaps on the trees and exactly described on a plat, whereon shall be noted by the Surveyor, at the proper distances, all mines, salt springs, salt licks and mill seats, that shall come to his knowledge, and all water courses mountains and other remarkable and permanent things over or near which such lines shall pass and also the quality of the lands.[2]

Here we find marked trees, the chain and the plat, already discussed as they appeared in the Jefferson report, carried into law.[3] Inserted into the paragraph were new descriptive obligations. In asking that mines, salt springs and licks, sites for mills, and the quality of the land be noted, the law assigned an exploratory mission to the surveyors. Formerly, as pointed out in the discussion of the Jefferson report, description was practically limited to cadastral purposes--that is, to notations which would assist in the identification of boundaries. Behind this expansion of descriptive duties may have stood, once again, the advice of Timothy Pickering. Pickering wrote, March 1, 1785, that he expected "the surveyors to be ordered to add to their surveys such explanations as would enable purchasers to judge of the value of the lands."[4]

To the identification of parcels of land the Ordinance brought the numerical simplicity which a uniform grid encouraged, and which the Jefferson report had hesitated to apply.[5] Townships were to be designated "by numbers progressively from south

---

[1] See section headed "Trees, Chain, Plat and Compass," chap. iii, above.

[2] Carter, Territorial Papers, II, 13. See also Jrnls. Cont. Cong., XXVIII, 376.

[3] See section headed "Trees, Chain, Plat and Compass," chap. iii, above.

[4] Pickering to Gerry, Philadelphia, March 1, 1785, in Pickering, Life of Timothy Pickering, I, 505.

[5] See section headed "The Identification of Lots," chap. iii, above.

to North, always beginning each range with number one."[1] The "ranges," or north-south columns of townships, were to be "distinguished by their progressive numbers to the westward."[2] This was plainly a scheme appropriate only to an area where lands to the south and east of certain bounding lines lay outside the field of survey. Its employment in the area which Congress had in mind--west of the Pennsylvania line and north of the Ohio River[3]--is shown in Figure 10A. The numbering of townships was not reduced to a universally applicable order until twenty years later, when a prime meridian and base line first divided a field of survey into quadrants, thus allowing the numbering of ranges both eastward and westward from an initial meridian, and the numbering of townships both northward and southward from a base line.[4] This innovation will be discussed in one of the closing chapters of this study.

Square-mile lots within the townships were to be "numbered from 1 to 36, always beginning the succeeding range of the lots with the number next to that with which the preceeding one concluded."[5] Under this rule, numbering could begin in any corner of a township. If the term "range" had the same meaning here as it had with reference to townships, then horizontal progression, as in the Jefferson report (Fig. 7B), was ruled out, but eight different specific orders remained possible. The order later chosen, presumably by the Board of Treasury, is shown on an accompanying diagram (Fig. 10B).[6]

---

[1] Carter, Territorial Papers, II, 13. See also Jrnls. Cont. Cong., XXVIII, 376.

[2] Carter, Territorial Papers, II, 13. See also Jrnls. Cont. Cong., XXVIII, 376.

[3] The surveys were to extend north to Lake Erie, as planned at this time. (Jrnls. Cont. Cong., XXVIII, 376.) Hence, Fig. 10A shows lines projecting well to the north of the area actually surveyed under the Ordinance. That the Ohio River was expected to serve as a continuous southern boundary was made evident in a motion in Congress respecting sales, ibid., p. 337.

[4] These master axes were laid out in southern Indiana, under the direction of Jared Mansfield, 1804.

[5] Carter, Territorial Papers, II, 13; Jrnls. Cont. Cong., XXVIII, 376.

[6] A square-mile grid appears on the face of thirty-six of the seventy-seven original plats prepared for the "Seven Ranges," now deposited in Records of the General Land Office (Record Group 49), Cartographic Records Branch, the National Archives. Each of

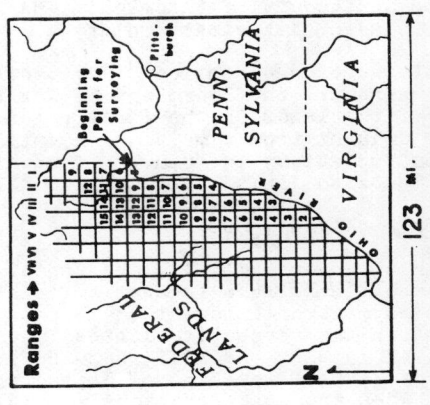

Fig. 10

Geographer and Surveyors

As indicated by Monroe, in the summary quoted near the beginning of this chapter, thirteen surveyors were to be appointed, to act under the control of a geographer. The sole Geographer, after the resignation of Simeon De Witt, in 1784, was Thomas Hutchins.[1] The idea of an office of register, which Hutchins had said "would undoubtedly be very acceptable,"[2] had been abandoned, along with the system of procuring land through warrants.[3] This change first appeared in the Grayson report, and on April 23, nearly a month before the passage of the Ordinance, it was ordered that Hutchins "be informed that Congress have occasion for his immediate services."[4]

The Geographer was to have no concern with sales, which were assigned to the "Commissioners of the loan office of the several states."[5] (Formerly, under the ordinance proposed by Jeffer-

---

the gridded plats displays the number-sequence shown in Fig. 10B. Neither gridding nor numbering, to judge by visual evidence, were features of the plats as submitted by the respective surveyors.

[1] See section headed "Registers and Surveyors," chap. iii, above.

[2] Hutchins to John Montgomery, Philadelphia, May 19, 1784, John Montgomery Papers, Chicago Historical Society.

[3] Military warrants were an exception. For the satisfaction of holders of such warrants, the Secretary of War was to take by lot one seventh of the lands in the first seven ranges surveyed, and "a similar draught from time to time until a sufficient quantity is drawn to satisfy the same." Carter, Territorial Papers, II, 14. See also Jrnls. Cont. Cong., XXVIII, 377.

[4] Jrnls. Cont. Cong., XXVIII, 291.

[5] Carter, Territorial Papers, II, 14, and Jrnls. Cont. Cong., XXVIII, 377. Loan officers were federal officials primarily responsible for the payment of interest on loans contracted by the federal government in the respective states in which they were stationed. (Ibid., XXVI, 312, and XXIX, 583-584.) For a full list of the duties of loan officers as of September, 1785, see ibid., pp. 792-794. When the Land Ordinance was being framed, the dispersion of sales among the states was apparently insisted upon by delegates to Congress. Grayson found that "the idea of allowing the Citizens of each State an equal chance of trying the good lands at their own doors," was one of the strongest reasons with them for consenting to the ordinance. (Grayson to Washington, New York, May [8], 1785, in Burnett, Letters of Members, VIII, 118.) Before any sales were held, however, Congress released the Board of Treasury from the requirement that lands be disposed of through the state loan officers. (Jrnls. Cont. Cong., XXXIV, 306-307.) The first and only auction under the Land Ordinance of 1785 was held at New York, September 21-October 9, 1787.

son's committee, these same officers were to sell, in the seaboard states, the land warrants made up by the Registers of the new western states.[1]) The Geographer was to direct the work of the surveyors, to "personally attend to the running of the first east and west line," and to "take the latitudes of the extremes of the first north and south line and of the mouths of the principal rivers."[2] Upon the completion of each group of seven ranges of townships, he was to transmit plats thereof to the Board of Treasury.[3] Hutchins was to be, in all but name, Surveyor General.[4]

Surveyors, by the Land Ordinance of 1785, were briefly cast in a role for which there was no exact precedent, and to which, in the later history of the surveys, there would be no return. Technically, their assignment was no more exacting than that given in the Jefferson report of the preceding year.[5] Financially, their position was simplified, in that they were to be paid not in fees, by the purchasers of land, but by the federal government, proportionate to the number of miles surveyed.[6] Their general terms of employment, which obliged them to pay the hire of survey crews and other field expenses,[7] resembled those governing the later deputy surveyors, who, up to 1910, carried surveying

---

[1] Boyd, *Papers of Thomas Jefferson*, VII, 142; and *Jrnls. Cont. Cong.*, XXVI, 326.

[2] Carter, *Territorial Papers*, II, 13. The Geographer was also to administer oaths to the surveyors. *Ibid.*, p. 12.

[3] *Ibid.*, p. 14.

[4] A distinction between the functions of Geographer and Surveyor General was emphasized by an applicant for the latter office, after Hutchins' death in 1789. "Whether Congress . . . may think proper to establish such an Officer as Geographer or not," wrote the applicant, "I shall not presume to Conjecter, but must suppose it will be necessary to appoint a Superintendent or Surveyor General for the Western Country which was done by the late Geographer. . . ." Dorsey Pentecost to President Washington, Winchester [Virginia], July 10, 1789, Applications for Office under President Washington, Manuscripts Division, Library of Congress.

[5] The terms for the running and marking of lines were almost identical with those earlier prescribed, as has been shown, above.

[6] "Each surveyor shall be allowed and paid at the rate of two dollars for every mile in length he shall run." Carter, *Territorial Papers*, II, 13.

[7] The surveyor's pay was to cover "the wages of chain carriers, markers and every other expence attending the same." *Ibid.*

over the greater part of the public domain, under the Land Act of 1796 and subsequent laws.[1] But surveyors under the Land Ordinance of 1785 were not bound by contract, as were the deputy surveyors. Most important, each of the original thirteen surveyors was to go forth as the special representative of his home state, nominated by his state's delegation in Congress.[2]

"The design of a surveyor from each State," wrote the New Hampshire delegates who had assisted in framing the Ordinance, "was that by going into the country they might be able to communicate information to the states for which they were appointed of the quality of the lands, and such other circumstances as may direct the citizens in making their purchases."[3] The Connecticut delegates pointed out to their governor that the man to be appointed from their state "sho'd not only be well skilled in the Art of Surveying but possess Talents for Observation and discovery." "It is of much importance," they continued, "that there sho'd be no disappointment in this regard by a declining of the Appointment . . . [to] undertake this important and arduous task."[4] In view of the qualifications expected, it should not be surprising to hear that the appointees who later assembled to survey under the Ordinance were styled, "the Gentlemen Surveyors."[5]

The Place of Beginning

Surveying was to begin, in the words of the Ordinance, "on the River Ohio at a point that shall be found to be due north from the western termination of a line which has been run as the

---

[1] For details of the role of deputy surveyor, see Lowell O. Stewart, Public Land Surveys, pp. 59-90.

[2] "A surveyor from each State shall be appointed by Congress or a committee of the States [i.e., a committee for carrying on government between the sessions of Congress]." Carter, Territorial Papers II, 12. Analogous to and connected with this provision was the arrangement for sales in each of the states, as noted above.

[3] New Hampshire Delegates to the Governor of New Hampshire, New York, May 29, 1785, in Burnett, Letters of Members, VIII, 130-131.

[4] Connecticut Delegates to Governor of Connecticut, New York, May 27, 1785, in Burnett, Letters of Members, VIII, 124-125.

[5] Hutchins to "The Gentlemen Surveyors," Geographer's Camp, July 21, 1786, Papers of the Continental Congress, LX, 249.

southern boundary of the state of Pennsylvania."[1] This was a considered designation, by which Congress was enabled to fix an initial point for the federal surveys without dictating Pennsylvania's western boundary.[2] Survey of Pennsylvania's southern boundary had been completed in 1784, and the boundary commissioners of Virginia and Pennsylvania had agreed to meet again in the middle of the following May, to run due north to the Ohio River, as Congress was aware.[3] With sixty-three miles to run, the commissioners could be expected to reach the Ohio without delaying the start of the public land surveys.

From this beginning point (Fig. 10A), surveyors of the national domain were to proceed westward. First to be surveyed would be the country of which George Washington had said, "This is the tract which, from local position and peculiar advantages, ought to be first settled in preference to any other whatever."[4] The geography of that country, as of 1785, will be the subject of the following chapter.

Review

Upon reading the Land Ordinance of 1785, Jefferson wrote

---

[1] Carter, *Territorial Papers*, II, 13.

[2] South of the Ohio River, Virginia and Pennsylvania were left to execute an agreement already reached, by running a line due north. North of the Ohio, it was assumed that this line would be continued as the boundary of a new state, but the issue was left "open to discussion hereafter." (Grayson to Washington, New York, May [8], 1785, in Burnett, *Letters of Members*, VIII, 118.) The Ordinance provided that "nothing herein shall be construed as fixing the western boundary of the State of Pennsylvania." (Carter, *Territorial Papers*, II, 13.)

[3] The southwestern corner of Pennsylvania had been marked, October 16, 1784. See Andrew Ellicott's entry in diary from that date, quoted in Catherine V. C. Mathews, *Andrew Ellicott, His Life and Letters* (New York: The Grafton Press, 1908), p. 23. Hutchins, who had served as one of the Pennsylvania commissioners in 1784, notified the President of Congress, in April, 1785, that the boundary commissioners were to assemble once more, May 16, 1785. See Hutchins to President of Congress, New York, April 21, 1785, Papers of the Continental Congress, LX, 181. Hutchins, in this letter, asked leave to continue as a boundary commissioner. Two days later, Congress resolved to inform him of their need for his services.

[4] Washington to President of Congress, Army Headquarters, Newburgh, New York, June 17, 1783, in Fitzpatrick, *Writings of George Washington*, XXVII, 17. That this area--Jefferson's State of Washington--would be first surveyed had been long expected. See Samuel Dick to Thomas Sinnickson, Annapolis, March 18, 1784, in Burnett, *Letters of Members*, VII, 473.

to Monroe, "I am much pleased with your ordinance, and think it improved from the first in the most material circumstances."[1] Adoption of the principle of survey prior to sale at public auction, as we have seen, was among the most important departures from the report of Jefferson's committee. The rectilinear grid was retained, but converted from a means of exerting control over the southern survey system to a framework for New England townships. Townships were to occupy tracts six statute miles square, and the opportunity for purchasing square-mile lots was to be confined to alternate tracts. Numbers became the sole means of identifying parcels of land. Marked trees, the plat, the chain and the compass all appeared in the final Ordinance much as they were handed on by Jefferson's committee. Surveyors were to act under the direction of a Geographer rather than Registers. They were now expected not only to survey but to explore, each on behalf of the state he represented. Finally, the entire Ordinance was directed toward the survey and disposal of lands in one specific area, extending from the Ohio River to Lake Erie, immediately west of Pennsylvania. Said Jefferson, who had undertaken legislation for the disposal of land in all of the western states authorized by Congress, "I had mistaken the object of the division of the lands among the states."[2]

---

[1] Jefferson to Monroe, Paris, August 28, 1785, in Boyd, **Papers of Thomas Jefferson**, XIII, 445.

[2] Ibid. Despite its shrunken area of reference, the final Land Ordinance retained the stature of a companion law to Jefferson's ordinance providing for the organization of western states (the Ordinance of 1784). These two laws were sent out from Congress late in May, 1785, as the joint expression of Congressional western policy. See Secretary of Congress to Certain States, May 28, 1785, in Burnett, **Letters of Members**, VIII, 128.

CHAPTER V

THE FIRST SCENE OF SURVEY

Work in the field, under the terms of the new Land Ordinance, began late in September, 1785. In spatial extent, the surveying which ensued was modest in the extreme, not even covering Jefferson's small "State of Washington." Work had barely begun when Congress limited the field of survey by a line running due west from the beginning point on the Ohio River. After the completion of seven ranges of townships south of this line, in 1787, Congress declined to authorize further progress under the original plan of survey. The first phase of U.S. public land surveying thereupon came to an end, although final returns were not submitted until July, 1788.

The Seven Ranges, as the townships laid out during this period have come to be termed, are undoubtedly better known by name than any of the land subdivisions later surveyed in the U.S. public domain.[1] The present chapter introduces the reader to the scene of survey, and the following three chapters undertake, respectively, to narrate events in the field, to discuss the major technical and financial problems which arose in the course of surveying, and to set forth an appreciation of the value of the work accomplished by the Geographer and surveyors, who engaged in this pioneer venture.

The purpose of this chapter is to develop a picture of only those elements in the scene of survey which bear upon the story of surveying which follows. Accordingly, five subjects are treated, in the order named: (1) Pittsburgh and the road from the East, (2) Fort McIntosh, (3) settlements on the upper Ohio River, (4) Indian tribes beyond the Seven Ranges, and (5) the lay of the land within the immediate field of survey.

---

[1] This is the only part of the public land survey system, for example, which is accorded separate attention in James Truslow Adams (ed.), Dictionary of American History (5 vols.; New York: Charles Scribner's Sons, 1942). See Eugene H. Roseboom, "Survey of the Seven Ranges," ibid., Vol. V.

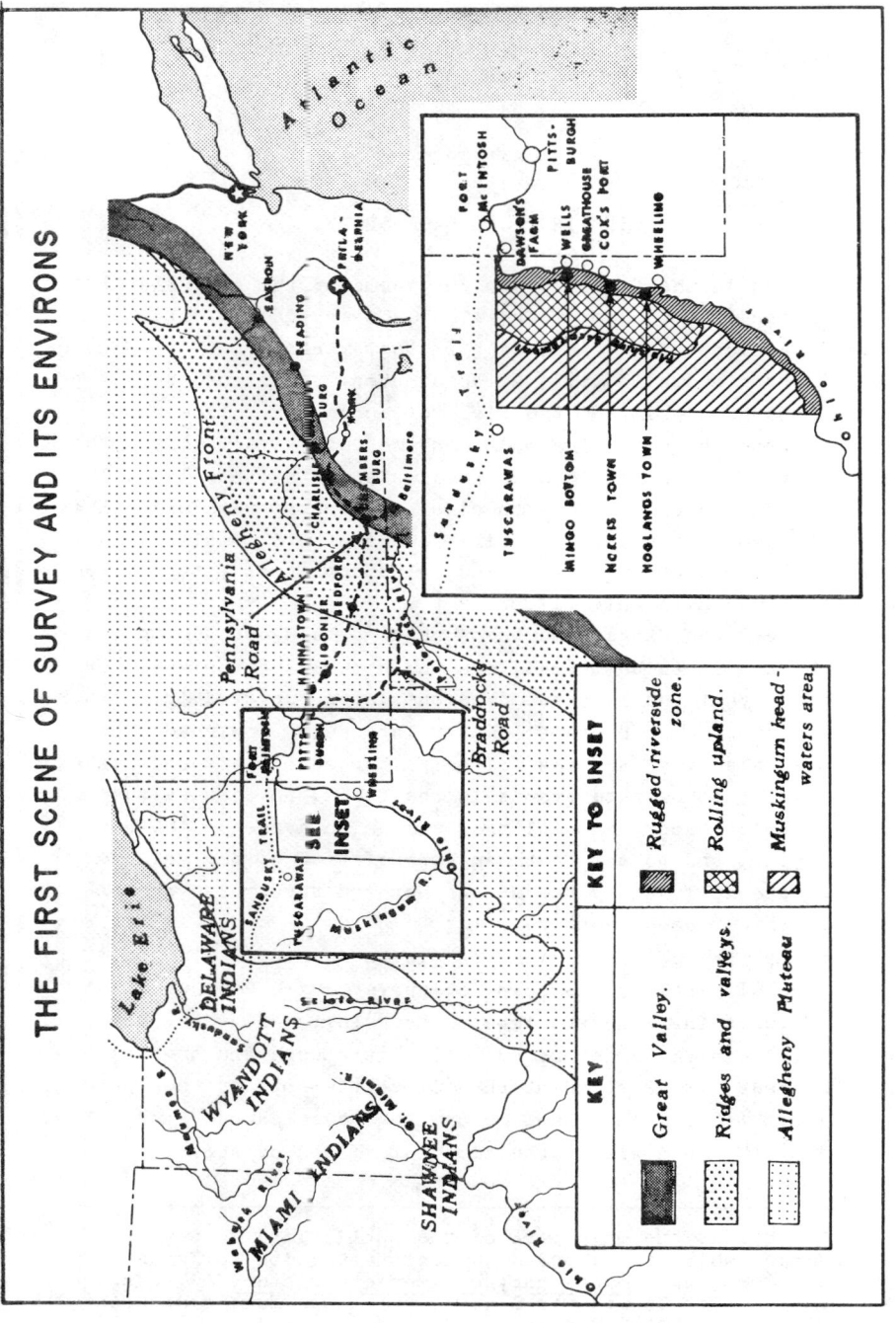

Fig. 11

Pittsburgh and the Road from the East

In speaking of the "local position and peculiar advantages" of the region of the Seven Ranges, George Washington was referring, in part, to the fact that this country lay along the Ohio River, not far downstream from Pittsburgh (Fig. 11). After thirty years of war-filled history as a fortified place, Pittsburgh was still a humble village of about three hundred persons, in 1785. Said Arthur Lee, who had paid a visit there in the preceding year:

> Its inhabitants are almost entirely Scotch and Irish, who live in paltry log cabins. . . . A great deal of small trade is carried on, mostly for barter, the goods being bought . . . from Philadelphia and Baltimore. They take in the shops money, wheat, flour and skins.[1]

Lee might also have noted that large eastern supply firms had representatives in the village. Pittsburgh was on the verge of becoming a major wholesaling and outfitting center for the West.

Pittsburgh was three hundred and twenty miles from Philadelphia by the "Pennsylvania Road," the more important of two main approaches from the East, in 1785 (Fig. 11).[2] From Philadelphia to York a regular stage service was available to the traveller, in a carriage which one of the surveyors found to be "without springs & very uneasy."[3] The trip took three days. A second stage brought the traveller into Carlisle, and from this point onward some private means of conveyance was required. Carlisle lay, surrounded by farms, in the Great Valley, a narrow tract of lowland which swept across the traveller's general westward course. This lowland corridor, known in southern Pennsylvania as the Cumberland Valley, and in Virginia as the Shenandoah Valley, was recognized even at this time as a continuous topographic feature "that extends itself from New Hampshire to

---

[1] "Arthur Lee's Journal," Olden Time, II (August, 1847), 339; entry for December 17, 1784.

[2] The other route, known as "Braddock's old road," led from Fort Cumberland on the Potomac River to a point on the Monongehela River near Pittsburgh (Fig. 11). This road, though more celebrated in song and story than the Pennsylvania Road, was much the less travelled of the two roads in the late eighteenth century.

[3] Diary of Winthrop Sargent, June 18 - December 21, 1786, Sargent Papers, Massachusetts Historical Society, Boston, Massachusetts; entry for June 30, 1786. The following description of travel on the Pennsylvania Road is based principally upon this diary.

Georgia between two of the Apalachian Mountains."[1] Entering the Great Valley at Easton, travellers from New York could reach Carlisle by way of Reading and Harrisburgh. From Carlisle, the road to Pittsburgh led down the Great Valley to Chambersburgh, joining there a road from Baltimore.

Beyond Chambersburgh, the main route of travel generally followed the northwesterly course of Forbes' Road, a military highway cut through to Pittsburgh by British forces in 1758.[2] Within a few miles of Chambersburgh, the road passed from the Great Valley, crossing the first of a series of long, even-crested, parallel ridges. A horseback rider from Carlisle could turn aside and leave the Great Valley by a convenient gap without travelling through Chambersburgh, and, rejoining the main road beyond a further ridge, could reach Bedford after three days' travel. Another half-day brought him to the Allegheny Front, where a way "very winding and by no means difficult of ascent" led him upward to "a large extent of land comparatively plain."[3] This soon gave way to what one surveyor called "some of the hilliest Country on earth."[4] The traveller had attained the Allegheny Plateau. Ahead, in the midst of the Plateau, more than two days' distant by way of Ligonier and Hannastown, lay Pittsburgh.

In the final part of the journey the main road, occasionally digressing into safer paths, left to the venturesome horseback rider "a Route where you are compelled to dismount or very much risk your neck."[5] In the last day's travel the country, though said to have been very ill-cultivated and settled, began to display the first signs of extensive agriculture seen since the itinerant's departure from the meadows and wheatlands of the Great Valley, and his entry into the forests of western Pennsyl-

---

[1] Ibid., entry for July 4, 1786.

[2] A well-documented account of the opening of this road appears in John D. Barnhart, Valley of Democracy (Bloomington, Indiana: Indiana University Press, 1953), pp. 26-27.

[3] Samuel Holden Parsons to William Samuel Johnson, October 2, 1786, in Charles S. Hall, Life and Letters of Samuel Holden Parsons (Binghamton, N.Y.: Otseningo Publishing Co., 1905), p. 490.

[4] Andrew Ellicott to his wife, tent near Beesontown, October 3, 1784, in Mathews, Andrew Ellicott, p. 20.

[5] Diary of Winthrop Sargent, entry for July 12, 1786.

vania. Ten days in all, at the best rate of progress, brought the traveller from Philadelphia to his destination.

Fort McIntosh

Congress, in designating the beginning point for the surveys, was aware not only of its accessibility from the East, and its convenient location relative to Pittsburgh, but also of its nearness to a fort which stood on a bluff overlooking the Ohio River, twenty-seven miles downstream from Pittsburgh. This was Fort McIntosh (Fig. 11), a stockaded outpost which had been built and then abandoned during the Revolutionary War.[1] Troops sent out to garrison the fort late in 1784 found that emigrants floating down the Ohio had "destroyed the gates, drawn all the nails from the roofs, taken off all the boards, and plundered it of every article," but by the time the Land Ordinance was passed by Congress the fort had been restored to "tolerable good order."[2]

It was at Fort McIntosh that, early in 1785, Indian tribes of the Ohio country signed away their title to the land about to be surveyed. In response to a call from Congress, Pennsylvania had mustered a militia force of over two hundred men to attend this treaty conference. They were under the command of Colonel Josiah Harmar, who established his headquarters at Fort McIntosh, and in doing so made of the fort a temporary center of American military activity. At the time the Land Ordinance was enacted, Harmar anticipated the arrival of militia from Connecticut, New York and New Jersey, and a consequent enlargement of his command to a force of about seven hundred men.[3]

During the summer of 1785, as the time for starting the surveys approached, Harmar began to lose, rather than gain, men. Most of his Pennsylvania troops departed upon the expiration of

---

[1] For a description and history of this fort, built as an outpost of Fort Pitt, see Louis E. Graham, "Fort McIntosh," Western Pennsylvania Historical Magazine, XV (January, 1932), 93-119.

[2] Colonel Josiah Harmar to John Dickinson, Governor of Pennsylvania, Fort McIntosh, February 8, 1785, in Consul W. Butterfield (ed.), Journal of Capt. Jonathan Heart . . . to Which Is Added the Dickinson-Harmar Correspondence of 1784-5 (Albany, New York: Joel Munsell's Sons, 1885), p. 48.

[3] For a record of congressional action, in requesting state militia as a substitute for a standing army (then considered dangerous to the liberties of a free people), see Jrnls. Cont. Cong., XXVII, 524, and XXVIII, 224, 240, 247, 435. For an account of the early stages in concentration of militia at the gateway to the Ohio country, see correspondence of Josiah Harmar, August, 1784-June, 1785, in Butterfield, op. cit., pp. 46-74.

their terms of enlistment, due to "their want of confidence in the public treasury respecting pay and their wish to go down the country [to their homes in the East]."[1] Only seventy men were induced to reënlist, but early in September a company of New York artillery arrived, and from this time onward the number of troops under Harmar's command continued to grow.[2]

One of the first acts of Thomas Hutchins, upon his arrival in Pittsburgh in September, 1785, was to proceed down the Ohio to Fort McIntosh to confer with Colonel Harmar. It was Harmar's advice, that Hutchins could "very safely repair with the Surveyors to the intersection of the West line of Pennsylvania with the Ohio," which emboldened the Geographer to go forward with his surveying plans.[3]

Settlements on the Upper Ohio River

Downstream from the point where the surveys were scheduled to begin, the Ohio River formed a legal frontier. It not only separated the national domain from territory under the jurisdiction of Virginia, but it also marked the limit of legal settlement, since Congress had banned all settlement on the public domain pending regular survey and sale of the land.[4] This was the first of a series of such temporary legal frontiers which for more than a century marked the progress of American expansion across the vast public domain. Here, as in later instances, the line of demarcation was violated by impatient "squatters." In 1785, settlers could be found on both sides of the Ohio, despite the law.

On the left bank, or Virginia shore, there were several very small settlements, where aid and comfort, already being dispensed to emigrants bound for Kentucky, awaited the federal surveyors (Fig. 11). Of these scattered clearings in the wilderness,

---

[1] Harmar to Dickinson, Fort McIntosh, September 1, 1785, ibid., p. 85.

[2] Harmar to General Knox, Fort McIntosh, October 22, 1785, ibid., pp. 92-93.

[3] Thomas Hutchins to President of Congress, Pittsburgh, September 15, 1785, Papers Cont. Cong., LX, 189.

[4] On September 22, 1783, Congress forbade all persons "from making settlements on lands inhabited or claimed by Indians, without the jurisdiction of any particular state." See Jrnls. Cont. Cong., XXV, 602. This ban first became effective beyond the Ohio River when Virginia ceded its jurisdiction there, in March, 1784.

some were single farms, such as that of a Mr. Dawson who was prepared to supply milk, butter and vegetables to the surveyors from his plot of land located near the beginning point of the surveys.[1] A few of the clearings along the river were organized into "stations," one of which was Cox's Fort, located about thirty miles downstream from Dawson's farm. Here, four or five log cabins were enclosed in a stockade, and other cabins stood in the immediate vicinity, forming altogether a community capable of defense against Indian attack.[2]

Four pioneer settlers of local renown, who were well established on the Virginia shore by the summer of 1785, later extended their hospitality and assistance to the surveyors: Charles Wells, at whose house, about ten miles above Cox's Fort, some of the surveyors wintered, 1786-87;[3] William Greathouse, located between Wells' and Cox's Fort, who came to the aid of the surveyors in the field;[4] William McMahon, a local magistrate, whose house at Cox's Fort was a frequent resort of the surveyors, and who arranged for the hire of both horses and field hands;[5] and Ebenezer Zane, who, as co-founder of the settlement already known at this date as Wheeling, was in a position to support the surveys in the same manner as McMahon.[6]

Settlers on the federal side or Indian shore seem to have differed from their counterparts across the Ohio in little more than the fugitive nature of their occupancy. They were generally

---

[1] Diary of Winthrop Sargent, entry for July 14, 1786.

[2] Ibid., entry for August 7, 1786. The importance of farms and stations along the upper Ohio River to early emigrants travelling down the river is recognized in Randolph C. Downes, Frontier Ohio, 1788-1803 (Columbus, Ohio: Ohio State Archaeological and Historical Society, 1935), p. 55.

[3] Journal of John Mathews, July 10, 1786-April 21, 1787, Marietta College Library, Marietta, Ohio; entry for December 26, 1786. Charles Wells, for whom Charleston, West Virginia, was named, later founded a settlement downstream from Wheeling.

[4] Ibid., entries for September 26 and 27, 1786.

[5] Ibid., entries for October 5, November 8, 14 and 15, and December 3, 1786; Diary of Winthrop Sargent, entry for November 10, 1786.

[6] Ibid., entries for July 21 and November 7, 1786. Ebenezer Zane, who with Jonathan Zane had founded Wheeling in 1769, was to figure importantly in the subsequent settlement of the Ohio country as operator of a ferry across the Ohio at Wheeling, and creator of "Zane's Trace," a road which led from a point opposite Wheeling westward to a point across the Ohio from Maysville (Limestone), Kentucky.

looked upon, however, as "banditti," and their advances into the Northwest caused widespread concern in official circles. As early as 1779, an officer sent down the Ohio from Pittsburgh found "small improvements all the way from the Muskingum River to Fort McIntosh & thirty miles up some of the Branches."[1] Feared during the Revolutionary War because of their provocative effect on the Indians, these aggressive settlers came to be viewed, in 1785, as a threat to the program for the orderly survey and sale of federal lands.[2]

There were three principal settlements on the federal side of the Ohio between the Pennsylvania line and a point opposite Wheeling (Fig. 11). Furthest upstream of these was a clearing of about two hundred acres at "Mingo Bottom," south of present-day Steubenville. The former site of a Seneca (Mingo) Indian village, this place was found to harbor a body of tenants under one Joseph Ross, when an officer from Fort McIntosh attempted to clear the country in advance of the surveyors, in April, 1785.[3] Despite a promise to remove themselves permanently, Ross and his followers were discovered on the same ground six months later.[4] About ten miles down the river, near the site of modern Warrentown, was "Norris Town," where the evicting officer found about forty men armed to resist removal and ready with a petition to Congress to grant them permission to remain.[5] These settlers, led by one Charles Norris, still held their ground after surveying had begun.[6] A few miles further downstream, near the site of modern

---

[1] Colonel Brodhead to General Washington, Fort Pitt, October 29, 1779, quoted in Randolph C. Downes, "Ohio's Squatter Governor: William Hogland of Hoglandstown," Ohio Archaeological and Historical Publications, XLIII (1934), 274-275.

[2] For expressions of this fear, see William Grayson to George Washington in Burnett, Letters of Members, VIII, 118; and "Arthur Lee's Journal," Olden Time, II (August, 1847), 340; entry for December 19, 1784.

[3] Ensign Armstrong to Colonel Harmar, Fort McIntosh, April 12, 1785, in Archer B. Hulbert (ed.), Ohio in the Time of the Confederation ("Marietta College Historical Collections, Ohio Company Series," Vol. III; Marietta, Ohio: Marietta Historical Commission, 1918), p. 107.

[4] "Journal of General Butler," Olden Time, II (October, 1847), 437; entry for October 1, 1785.

[5] Armstrong to Harmar, Fort McIntosh, April 12, 1785, in Hulbert, Ohio in Time of Confederation, p. 108; and "Petition of Inhabitants West of the Ohio River," ibid., p. 105.

[6] "Journal of General Butler," Olden Time, II, 438; entry for October 2, 1785.

Martin's Ferry, stood the cabins of "Hoglands Town," where the officer from Fort McIntosh gained, again, no more than promises of removal.[1] Despite the destruction of their property by later expeditions of federal troops, pioneers from this town were found on the opposite shore, in 1786, "furnished with flats to pass and repass themselves and effects across the river occasionally, as the movement of the troops may make it necessary."[2]

Although the surveyors did not enjoy hospitality in the cabins of the three settlements just described, they found welcome shelter on several occasions in deserted cabins up-country from the Ohio River.[3] The squatter population on the Federal side of the Ohio may have furnished some of the hunters and field hands who were recruited for the survey parties.

Indian Tribes

The region of the Seven Ranges had already ceased to be Indian country, when surveying began. While it is true, as we shall see, that the surveyors had direct encounters with Indians in the course of their work, these were small bands of itinerant savages, whom they met. The serious threats of organized Indian resistance, which repeatedly delayed the progress of surveying, originated among tribes residing well beyond the immediate field of survey. The four principal tribes concerned will be passed in review here, following a note on the British, who stood behind them.

The key to threatened Indian resistance, as the Geographer Thomas Hutchins realized soon after his arrival in the West, was Detroit.[4] The British remained in possession of Detroit, despite the terms of peace which concluded the Revolutionary War, and they displayed no intention of relinquishing this outpost and

---

[1] Armstrong to Harmar, Fort McIntosh, April 12, 1785, in Hulbert, Ohio in Time of Confederation, p. 108. For full interpretation of the significance of a local election, news of which Enisn Armstrong learned at this place, see Downes, "Ohio's Squatter Governor," Ohio Archaeological and Historical Publications, XLIII (1934), 273-282.

[2] Diary of Winthrop Sargent, entry for August 7, 1786.

[3] Ibid., entry for September 24, 1786; and Journal of Thomas Hutchins, entries for August 8 and 15, 1786, Hutchins Papers, II, 93.

[4] Hutchins to President of Congress, Pittsburgh, September 15, 1785, Papers Cont. Cong., LX, 189-191.

others to American troops.[1] From Detroit, the British maintained a fur trade upon which the Indians depended, encouraged unity among the tribes, and propagated the proposition that, although they had by treaty of peace "given the Americans the Law or Jurisdiction over the Country, . . . they did not give them any right to the Land."[2] Staying in the background, the British seem to have upheld the Indian cause in the Northwest to this extent and no further, reportedly saying to the Indians that "if the Americans wanted Land from them they and the Americans must agree on that matter, as the British had nothing now to do with it."[3]

Of the four principal tribes who stood between the British at Detroit and the advancing American frontier along the Ohio River, two had sent chiefs to sign the Treaty of Fort McIntosh.[4] These two tribes, who were later expected by Hutchins to send chiefs to guarantee the safety of the surveyors, were the Delaware and Wyandott. The Delaware, whose original home was on the eastern seaboard, as their name suggests, had retired to the Ohio country, where they commanded the entire Muskingum valley until shortly before the appearance of the government surveyors. This tribe, numbering perhaps four hundred persons, now claimed only the upper waters of the Muskingum, and nearby areas to the north and west, all approximately contained within the cession line of the Treaty of Fort McIntosh. Their foremost salient, at "Tuscarawas," lay about fifty miles west of the beginning point of the surveys (Fig. 11).[5]

The Wyandott could be reached by following a major Indian

---

[1] For discussion of continued British occupation of posts at Oswego, Ogdensburg, Mackinac and Detroit, see A. L. Burt, "A New Approach to the Problem of Western Posts," *Report of the Annual Meeting of the Canadian Historical Association* (Ottawa, 1931), pp. 61-75.

[2] Deposition of William Wilson, Hutchins to President of Congress, New York, November 24, 1785, Enclosure No. 4, Papers Cont. Cong., LX, 217-220.

[3] Ibid.

[4] The signing of the Treaty of Fort McIntosh, January 21, 1785, has been noted in chap. 1, under "Indian Cessions."

[5] For an account of the migrations and troubles of the Delaware Indians, see Downes, Council Fires, pp. 41, 118, 121-122, 168-169, 263-265, 292-294. This tribe is principally remembered today because of a massacre of a group of Christian Delaware by frontiersmen, in 1782, on which see Beverley W. Bond, Jr., *The Foundations of Ohio*, Vol. I of *The State of Ohio*, ed. Carl Wittke (6 vols.; Columbus, Ohio: Ohio State Archaeological and Historical Society, 1941-1942), pp. 231-232.

115

thoroughfare, known as the Sandusky Trail, which led westward from the Ohio River, across the breadth of the Seven Ranges, and into the Delaware country, beginning at "Tuscarawas" (Fig. 11).[1] The Wyandott, less numerous than the Delaware, commanded the valley of the Sandusky River, which provided them with easy access to Detroit by way of Lake Erie. While closely associated with the Delaware, this tribe tended to be more receptive to British influence, and much more apprehensive of incurring the displeasure of other Indian tribes further west.[2] A Wyandott chief described his position as one "between two fires," explaining in a message to the Americans, "I am afraid of you, and likewise of the back Nations."[3]

The two most important "back nations" were the Shawnee and the Miami, both sworn enemies of the Americans. The Shawnee, numbering about the same as the Wyandott, commanded tributaries of the Ohio River in the southwest quarter of present-day Ohio (Fig. 11). For over a decade they had engaged in vengeful warfare with Kentucky pioneers.[4] Learning of plans for the federal surveys, a Shawnee chief protested to American emissaries, "We do not understand measuring out the lands--it is all ours. . . . Brothers, you seem to grow proud because you have thrown down the King of England."[5]

The Miami Indians, native to the upper Wabash River, had extended their settlements eastward into present-day Ohio before the Revolutionary War.[6] Even so, they remained relatively remote

---

[1] This trail, which formed a connection between Pittsburgh and Detroit in its full extent, is called "the most important trail in the central west" in Archer B. Hulbert, "The Indian Thoroughfares of Ohio," Ohio Archaeological and Historical Publications, VIII (1900), 263-295.

[2] On the relation of the Wyandott to other tribes, and to the British, see Downes, Council Fires, pp. 191-193, 239-241, 282-283.

[3] Transcribed speech of Captain Pipe, Jacob Springer to Thomas Hutchins, 38 miles on the East and West Line, September 13, 1786, Papers Cont. Cong., LX, 258.

[4] On this warfare, see Bond, Foundations of Ohio, pp. 224-229, 235-236. For earlier migrations and battles of the Shawnee, see Downes, Council Fires, pp. 41, 46, 163, 174, 177.

[5] Speech of Chief Kekewepellethy at Fort Finney, February 7, 1786, in "General Butler's Journal," Olden Time, II (November, 1847), 522; entry for January 30, 1786.

[6] Downes, Council Fires, p. 46. The Miami comprised six

from the immediate field of American expansion, in 1785 (Fig. 11). From their momentarily secure position, the Miami maintained an undisguised alliance with the British at Detroit, and sought to rally the more exposed tribes to defy the Americans. After the Treaty of Fort McIntosh, the Miami were hosts to a general Indian conference, intended to "brighten the chain of friendship" among the tribes of the Northwest, and to lead to the undoing of the treaty which had sanctioned the advent of surveyors on the public domain.[1]

It is hardly surprising that the Delaware and Wyandott disappointed Hutchins' expectations that they would follow up their capitulation at Fort McIntosh by sending chiefs to guarantee the safety of the surveyors. Wavering in their policy, they seemed to justify the apprehensions of an officer at Fort McIntosh who wrote, in September, 1785, "We shall have trouble in this country ere long, unless something is done . . . to avert the storm."[2]

Lay of the Land in the Seven Ranges

No picture of the scene of survey would be complete, of course, without a description of the land that the surveyors traversed. Of primary importance is the fact that the area of the first surveys lay entirely within the Allegheny Plateau (Fig. 11), a broad tract of hilly country representing the stream-carved remains of a plateau surface which, in the geologic past, sloped gently westward from the Appalachian Mountains.[3] The surveyors made their first acquaintance with this region, as the reader will

---

distinct tribes, of which the Wea and the Piankashaw were the most important.

[1] Samuel Montgomery, an American messenger who visited all four of the major tribes described here in August and September of 1785, found the Miami, then intriguing for a general renunciation of the Treaty of Fort McIntosh, "hostile and ill-disposed, blinded and misguided by the British." David I. Bushnell (ed.), "Journal of Samuel Montgomery," Mississippi Valley Historical Review, II (September, 1915), 271.

[2] Captain Doughty to General Knox, October 21, 1785, in Butterfield, Journal of Jonathan Heart, p. 90.

[3] The unglaciated part of the Allegheny Plateau, which covers much of western Pennsylvania, West Virginia, eastern Ohio and eastern Kentucky, is termed by Nevin M. Fenneman the "largest and most typical" section of the Appalachian Plateau Province, in his Physiography of the Eastern United States (New York: McGraw-Hill Book Company, 1938), p. 283. For genetic description of the area, see ibid., pp. 290-304.

recall, when they climbed the Allegheny Front, on the road to Pittsburgh. From that point onward, the face of the West as they knew it was that of the Allegheny Plateau.

As one examines the lay of the land within the Seven Ranges today,[1] three types of countryside emerge (Fig. 11). First, there is the rugged terrain bordering the channel which the Ohio River has cut into the plateau. Here, a series of short streams, seeking the level of the Ohio, have produced a countryside of steep slopes and scant upland surface, where hills rise five hundred or more feet above surrounding valleys.[2] Second, there are upland tracts lying ten to twenty miles from the Ohio, where moderate, rolling surfaces appear among the headwaters of these same streams.[3] Third, there is the countryside west of the Flushing Escarpment, a conspicuous landscape feature which generally divides the drainage of direct tributaries of the Ohio from streams which flow toward the Muskingum River. Here, in the upper reaches of the valley of the Muskingum, ridges are narrower, indentations are deeper, and small streams are more numerous than are those which are to be found east of the divide. Not so rugged as the district bordering the Ohio, this area is distinguishable from its immediate neighbor even to the eye of the casual traveller.[4]

Since George Washington so strongly recommended this general area for early settlement, one naturally wonders about his opinion of the land itself, as distinct from the apparent advantages of its location. He had viewed the country at first hand, from the Ohio River, and he was not deceived.[5] He was most

---

[1] The present account is based in part upon a first-hand acquaintance with the area, gained in September, 1955, and June, 1956.

[2] The local relief here averages between six hundred and eight hundred feet. See Guy-Harold Smith, "The Relative Relief of Ohio," Geographical Review, XXV (April, 1935), 277.

[3] These tracts have been identified as the somewhat lowered remnants of an erosion surface called the Harrisburg peneplain. Wilbur Stout and G. F. Lamb, "Physiographic Features of Southeast Ohio," Ohio Journal of Science, XXXVIII (March, 1938), 4.

[4] Along the Flushing Escarpment, the Harrisburg peneplain breaks down to the lower Lexington level. (Ibid., pp. 4-5.) For development of the contrast between the areas east and west of the Flushing Escarpment, see ibid., pp. 6-8.

[5] Washington, in the autumn of 1770, explored the course of the Ohio River from Fort Pitt to the mouth of the Kanawha River, stopping at points identified by Guy-Harold Smith, in his

attracted by the "bottoms," or alluvial lands along the Ohio and the lower reaches of its tributaries. Judging the fertility of these discrete, narrow tracts by their vegetation, he acknowledged that even these lands, the richest in the region, had an important drawback: their vulnerability to floods.[1] He recognized that the hills bordering the Ohio River were "not a range of Hills, but broken and cut . . . as though there were frequent water courses running through,"[2] and he considered their slopes so steep as to render them only "fit to support the Bottoms with Timber and Wood."[3] Further, he correctly guessed that extensive areas of relatively level land would not be found "till one gets far enough from the River to head the little runs and drains that come through the Hills, and to the Sources (or near it) of the Creeks and there Branches."[4] Lastly, he appreciated the fact that the most promising lands of the West lay not in this convenient locality, but in the great Central Lowland, beyond the limits of the Allegheny Plateau.[5]

It should be remarked, in conclusion, that the lay of the land within the Seven Ranges has in the long run won out over the surveyed grid, as a determinant of the "pattern of occupance." Although the original townships have been subdivided into small tracts in a standard, rectilinear fashion, the resulting lines of survey have generally failed of perpetuation, save as property boundaries. A view of the Seven Ranges from the air shows fencelines, roads and buildings arranged in general conformity with the terrain, usually to the disregard of the surveyed grid.[6] It is ironic that in this cradle, so to speak, of the rectangular surveying system, an influence which has shaped the greater part of the entire American landscape should have yielded almost completely to the dictates of Nature.

"Washington's Camp Sites on the Ohio River," *Ohio Archaeological and Historical Quarterly*, XI (January, 1932), 1-19.

[1] John C. Fitzpatrick (ed.), *The Diaries of George Washington, 1748-1799* (4 vols.; Boston: Houghton Mifflin Company, 1925), I, 442; entry for November 17, 1770.

[2] *Ibid.*, p. 415; entry for October 22, 1770.

[3] *Ibid.*, p. 442; entry for November 17, 1770.

[4] *Ibid.*   [5] *Ibid.*

[6] This assertion is based upon notes taken by the author during a trip by airplane over the area of the Seven Ranges, June 25, 1956.

CHAPTER VI

A CHRONICLE OF SURVEYING, 1785-1788

The following pages present chronologically an account of the survey of the Seven Ranges. There was further surveying, beyond the Seven Ranges, under the Land Ordinance of 1785, but this work was executed under private auspices, within the confines of two large land grants. It will be described in a later chapter. The present chapter covers all of the township surveying undertaken by the national government before the passage of the Land Act of 1796.

Establishment of the Beginning Point

The Land Ordinance was exactly three months old on the day that the beginning point of the public land surveys was established by boundary commissioners representing the states of Virginia and Pennsylvania. Setting off from the southwest corner of Pennsylvania in the first week of June, 1785, four commissioners led a party of perhaps thirty men in the survey of a line due north to the Ohio River (Fig. 12), where they arrived August 20.[1] Junction with the south shore of the Ohio completed the work of separating the territory of Virginia from that of Pennsylvania, but the commissioners sent field hands across the Ohio on that same day, to "set a stake on the flat, the North Side of the River."[2] At this stake, public land surveying would soon begin.

Of the four commissioners concerned in this boundary survey, two are worthy of special note: David Rittenhouse, representing Pennsylvania, who was a leading American scientist and at

---

[1]For coverage of period of survey, see journal and letters of Andrew Ellicott in Mathews, Andrew Ellicott, pp. 40-46. Establishment of the southwest corner of Pennsylvania by extension of the Mason and Dixon Line, in 1784, has been noted in chap. i of this study, under "Jefferson's Plan for Western States."

[2]Entry of August 20, 1785, in Journal of Andrew Porter, reproduced in William A. Porter, "A Sketch of the Life of General Andrew Porter," Pennsylvania Magazine of History and Biography, IV (1880), 268.

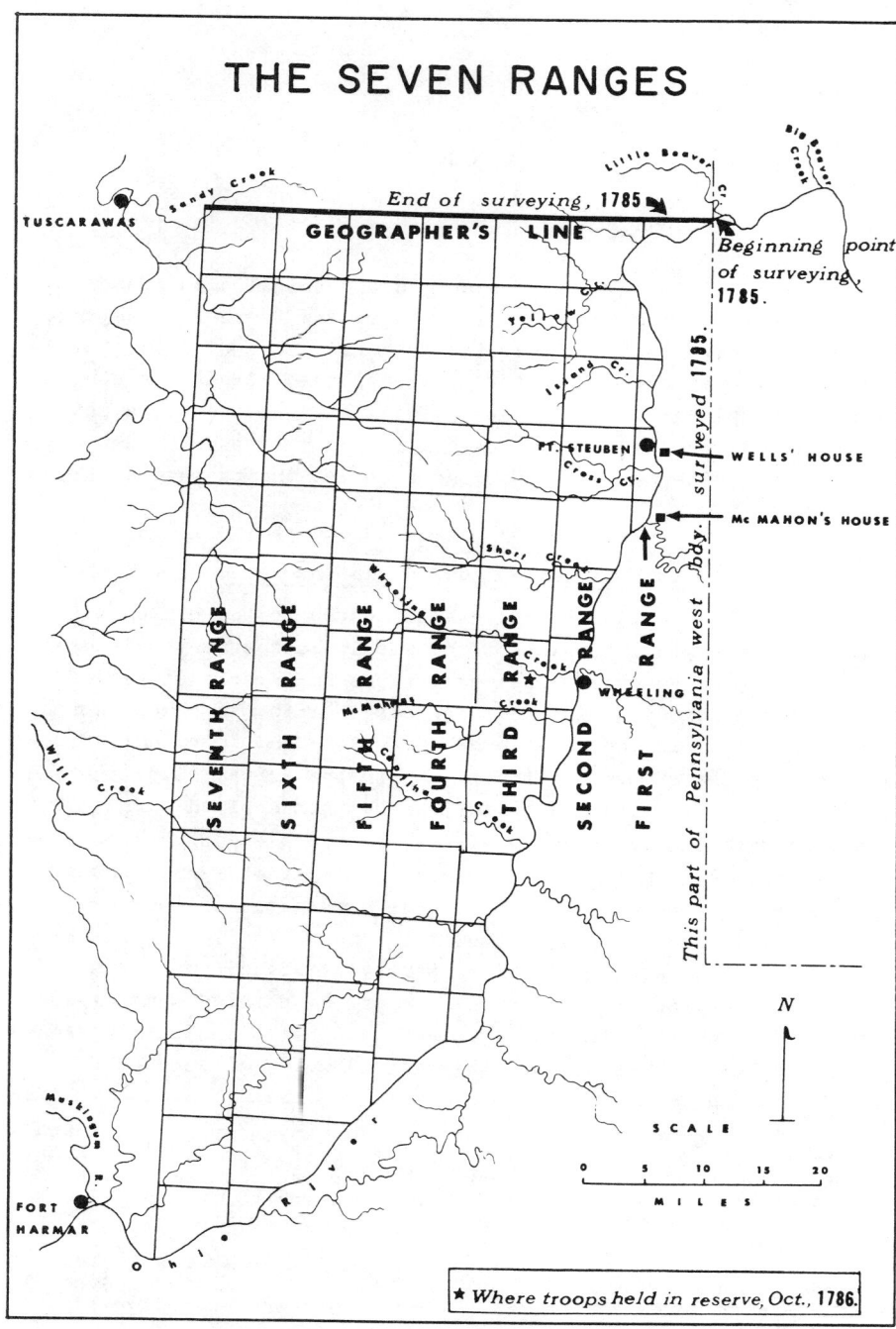

Fig. 12

this time the State Treasurer of Pennsylvania;[1] and Andrew Ellicott, representing Virginia, who later gained attention for his part in the design of the streets of Washington, D.C., and for his survey of the boundary between the United States and Spanish West Florida.[2] These two men, with their respective colleagues,[3] directed a corps of axemen who marked the passage of the boundary through the forests of the Allegheny Plateau by "cutting a wide vista over all the principal hills . . . and by falling or deadening a line of trees generally through all the lower ground."[4] The boundary itself was aligned through the use of a "Transit Instrument," equipped with what Ellicott termed "a most excellent Telescope."[5]

The transit instrument was set at successive stations, from one-half mile to two miles apart, upon ridge-tops.[6] Orientation of the instrument to true north was maintained by observations at night of the pole star (Polaris) and other stars, as well.[7] Demarcation of the boundary represented surveying of the highest quality known in the United States at this time. Said Jefferson, with reference to the boundary, "What is done by Rittenhouse can be better done by no one."[8] Errors were later discovered

---

[1] For an account of the life of Rittenhouse, see Edward Ford, *David Rittenhouse, Astronomer-Patriot, 1732-1796* (Philadelphia: University of Pennsylvania Press, 1946).

[2] On Ellicott's survey of the boundaries of the District of Columbia and his modification of L'Enfant's plan for the City of Washington, see Mathews, *Andrew Ellicott*, pp. 81-103. On Ellicott's expedition, 1796-1800, undertaken for the survey of the thirty-first parallel, which bounded the United States on the south by the treaty of peace ending the American Revolution, see *ibid.*, pp. 125-197.

[3] For Pennsylvania, Andrew Porter; for Virginia, Joseph Neville.

[4] Report of boundary commissioners, quoted in Mathews, *Andrew Ellicott*, p. 46.

[5] Ellicott to his wife, Wheeling Creek, July 6, 1785, *ibid.*, pp. 44-45. The use of the transit, a "meridional instrument" designed for observations in a vertical plane, was later described by Ellicott in a letter to Robert Patterson, Philadelphia, September 23, 1800, reproduced in Andrew Ellicott, *The Journal of Andrew Ellicott, Late Commissioner . . . for Determining the Boundary between the United States, and the Possessions of His Catholic Majesty in America* (Philadelphia, 1814), Appendix, pp. 51-52.

[6] Porter, "Life of Porter," *Pennsylvania Magazine of History and Biography*, IV, 275-276; journal entries for July 30-August 7, 1785.

[7] *Ibid.*, journal entry for June 9, 1786.

[8] Thomas Jefferson to Wilson C. Nicholas, Poplar Forest,

in the sixty-three mile course of the line, but many years were to pass before the standard of accuracy set here would be matched by the federal public land surveys.[1]

A few days after Pennsylvania's western boundary had been brought to the Ohio River, its extension further north was left in the hands of Andrew Ellicott, now acting on behalf of Pennsylvania, and a second commissioner, Andrew Porter. Although the line from the Ohio River to Lake Erie has come to be known as "Ellicott's Line," Ellicott joined in surveying only a part of it, in 1785.[2] It was completed in 1786 by Porter and another commissioner.[3] By the time the boundary reached Lake Erie, of course, the federal public land surveys were well under way.

The Geographer and Surveyors Assemble

The summer of 1785 was nearly gone when the Geographer, Thomas Hutchins, arrived in Pittsburgh from New York, where he had several months earlier placed himself at the disposal of the Continental Congress.[4] It was on September 4, the day after his arrival, that he consulted with Colonel Harmar at Fort McIntosh, receiving from that officer the assurance that he could "very safely repair with the Surveyors to the intersection of the West line of Pennsylvania with the Ohio." Returning to Pittsburgh, Hutchins joined several surveyors who had been in the village for a week or more, in "engaging Chain Carriers, purchasing provisions,

---

April 19, 1816, in Lipscomb and Berg (eds.), Writings of Jefferson, XIII, 481.

[1] At the time of the line's running, it was believed to vary "but 1/10 of an inch in forty miles." ("General Butler's Journal," Olden Time, II [October, 1847], 436; entry for September 30, 1785.) It was later found to be about fifty feet off course, at the end of its sixty-three mile length. Report of C. H. Van Orden, February 9, 1784, in Report of Secretary of Internal Affairs of Pennsylvania (1887), p. 403.

[2] Ellicott left the line at about the half-way point in its course between the Ohio River and Lake Erie, to return to Philadelphia. Mathews, Andrew Ellicott, p. 49.

[3] Archibald McLean assisted Andrew Porter after Ellicott's departure in 1785. These two men reached Lake Erie September, 1786. (Porter, "Life of Porter," Pennsylvania Magazine of History and Biography, IV, 284; journal entry for September 14, 1786.)

[4] Hutchins, it will be recalled, was summoned by Congress in April, 1785, as the Land Ordinance neared passage. See above, chap. iv, "Geographer and Surveyors."

and Buying Horses &c." A general movement down the Ohio to an encampment at the mouth of the Little Beaver Creek (Fig. 12), near the scheduled initial point, began September 20.[1]

Hutchins was distinguished from the surveyors who accompanied him down the Ohio by more than the fact that he was head of the "Geographer's Department."[2] Earlier in his life, as a British officer, he had served at Fort Pitt, and had undertaken exploratory expeditions from that point northward to Lake Erie, overland to Lake Michigan and the upper Wabash Valley, and down the Ohio River to the Mississippi.[3] His general map of the West, compiled largely on the basis of these expeditions, had established him as an authority on the area.[4] The map served in the present instance as the surveyors' guide to the country they were about to enter.[5]

Thirteen surveyors, one from each of the original states

---

[1] Hutchins to President of Congress, Pittsburgh, September 15, 1785, and New York, November 24, 1785, Papers Cont. Cong., LX, 189-191 and 193-200, respectively.

[2] In 1785, the Geographer's Department was classed as one of the principal executive agencies of government, along with the Indian Department, Foreign Affairs Department, Military and Marine Department, and others. ("General Account of Receipts and Expenditures," Papers Cont. Cong., CXLI, Pt. II, 317-318.)

[3] Hutchins, born in Monmouth County, New Jersey, in 1730, was an officer in the British colonial service from 1756 to 1778. After serving at Fort Pitt, Fort Chartres, and Pensacola, Florida, he sailed for England in 1777. In 1778, he deserted the British service, departing secretly from London for Passy, France, where Benjamin Franklin furnished him with a letter of introduction to the Continental Congress. He then returned to America. In 1781, Congress appointed Hutchins Geographer to the Southern Army, and then Geographer to the United States, jointly with Simeon De Witt. (Anna M. Quattrocchi, "Thomas Hutchins, 1730-1789," pp. 1-208.) Hutchins received a new commission as Geographer, to run for three years, May 27, 1785. (Jrnls. Cont. Cong., XXVIII, 398.)

[4] Hutchins' map of the West, _A New Map of the Western Part of Virginia, Pennsylvania, Maryland and North Carolina_, as well as his descriptive text prepared to accompany it, and an extended biographical note (later superseded by Miss Quattrocchi's dissertation) may be found in Thomas Hutchins, _A Topographical Description of Virginia, Pennsylvania, Maryland and North Carolina_, ed. Frederick C. Hicks (Cleveland: The Burrows Brothers Company, 1904). Hutchins' map, according to the notes of the present author, had already been relied upon by such men as George Washington, Thomas Jefferson, Rufus Putnam and Arthur Lee. It was later used by Andrew Ellicott, Arthur St. Clair, and John Cleves Symmes, among others, as the best available general source on the area embraced by present-day Kentucky, Tennessee, Illinois, Indiana, Ohio and southern Michigan.

[5] Journal of Winthrop Sargent, entry for October 28, 1786.

of the United States, accepted appointments to serve under the Geographer, in response to invitations sent out by Congress during the summer of 1785. But of these, one fell ill, one stayed at home due probably to a well-founded doubt that his services would be needed that year, and three failed to appear for reasons unknown.[1] These were the eight surveyors who reported for duty in the West:

For New Hampshire, Edward Dowse. Originally, New Hampshire's delegates to Congress had expected to send west a man from their own state, of a public character, whose report on "the quality of the lands, and . . . other circumstances" would carry the stamp of authority and lead to the purchase of townships by associations of New Hampshire citizens.[2] Dowse, an obscure surveyor who had been "lately in the western country," was neither a native of New Hampshire nor known there, but he was available in New York when the delegates were forced to give up hope that a suitable man from their home state could be induced to go "so far abroad."[3]

For Massachusetts, Benjamin Tupper. The initial nomination to the surveyorship for Massachusetts was almost a foregone conclusion. The honor went to Rufus Putnam, who, as earlier mentioned in this study, had championed the idea of a grant of land in the general area of the Seven Ranges for the satisfaction of soldier bounties.[4] He accepted the appointment by Congress

---

[1] The five appointees who failed to appear for service in the West were Caleb Harris of Rhode Island, Adam Hoops of Pennsylvania, Mark McCall of Delaware, Absalom Tatom of North Carolina, and William Tate of South Carolina. Notices of their election appear in Jrnls. Cont. Cong., XXVIII, 398, and XXIX, 539-540. Their letters of acceptance are in Papers Cont. Cong., LXXVIII, Pt. XII, 403; Pt. XVI, 459; Pt. XVIII, 561; Pt. XXII, 305-306. Harris fell ill. (Ibid., Pt. XII, p. 356.) Hoops was apparently apprehensive that no surveying would occur. (Hoops to Hutchins, Philadelphia, April 30, 1786, Hutchins Papers, Vol. III.) The failure of McCall, Tatom, and Tate to appear for service remains unaccounted for.

[2] New Hampshire Delegates to President of New Hampshire, New York, May 29, 1785, in Burnett, Letters of Members, VIII, 130-131.

[3] Dowse was recommended for the surveyorship in letter of New Hampshire Delegates to President of New Hampshire, June 27, 1785, ibid., p. 153. For notice of his election, see Jrnls. Cont. Cong., XXIX, 654. He was accepted only after Nathaniel Adams and Ebenezer Sullivan, both prominent citizens of New Hampshire, declined the appointment. See ibid., and Papers Cont. Cong., LXXVII, Pt. I, 461, and New Hampshire Delegates to President of New Hampshire, New York, July 24, 1785, in Burnett, Letters of Members, VIII, 169.

[4] Putnam became and remained the official appointee for

immediately, explaining that his principal motive was "a wish to promote emigration from among my friends into that country." Due to a prior commitment, however, Putnam sent as substitute a friend and fellow general of the Continental Army, Benjamin Tupper.[1] Massachusetts' delegates found in Tupper, no less than they would have found in Putnam, a man of the type sought for in vain by the delegates from New Hampshire.[2] It later developed that Tupper's role was that of advance scout for the Ohio Company of Associates, a group of adventurers drawn mainly from Massachusetts, who within a few years assumed leadership in the settlement of the Ohio country.

For Connecticut, Isaac Sherman. The first choice of the Connecticut delegates was, again, a leading citizen of the state: Samuel Holden Parsons, who was known to have "long entertained ideas of establishing himself or at least finding an estate in . . . [the Ohio] country."[3] But Parsons discovered another means of gaining a first-hand knowledge of the West, and the appointment went to Colonel Isaac Sherman, son of the influential Roger Sherman, to whose skillful politicing the success of Connecticut's claim to the Western Reserve has been attributed.[4] The young

---

Massachusetts. (Jrnls. Cont. Cong., XXVIII, 398, and XXIX, 542.) Putnam had led the signers of the Newburgh petition, in 1783, in calling upon Congress to assign land to veterans of the Revolution. See above, chap. i, "Jefferson's Plan for Western States."

[1] Putnam to Congress, Boston, June 11, 1785, in Papers Cont. Cong., LVI, 161, and Rufus King to Henry Knox, New York, June 27, 1785, in Burnett, Letters of Members, VIII, 153. Putnam had already accepted an appointment as Surveyor General for Massachusetts lands in the District of Maine for 1785.

[2] Tupper, having served as an officer throughout the American Revolution, emerged from the war with the rank of Brigadier-General. (Francis B. Heitman, Historical Register of Officers of the Continental Army during the War of the Revolution [Washington, D.C.: The Rare Book Shop Publishing Company, Inc., 1914], p. 551.) He was known at this time to be closely associated with Putnam as a leader in planning for western colonization. (Putnam to Washington, Rutland, Massachusetts, April 5, 1784, in Rowena Buell [ed.], The Memoirs of Rufus Putnam and Certain Official Papers and Correspondence [Boston: Houghton, Mifflin and Company, 1903], pp. 223-225.)

[3] Parsons to William Samuel Johnson, Middleton, Connecticut, May 21, 1785, in Hall, Samuel Holden Parsons, pp. 464-466.

[4] Parsons was soon appointed one of three commissioners to treat with the Indians who had not attended the treaty conference of Fort McIntosh, and so found himself, in November, 1785, floating down the Ohio River and observing the land. (Ibid., pp. 470, 476.) Isaac Sherman had retired from the Continental Army in

Sherman may not have gone west primarily to gather intelligence concerning the Western Reserve lands, north of the Seven Ranges, but before completing his tour of duty as a federal surveyor he addressed a letter on this subject to the governor of Connecticut.[1]

For New Jersey, Absalom Martin. The eagerness of prominent men in New Jersey, including the governor of the state, to place a surveyor in the Ohio country, was indicated by the fact that solicitations on behalf of Captain Martin were forwarded to New York even before the passage of the Land Ordinance.[2] Martin undoubtedly went west as an agent of prospective investors, but the success with which he performed his mission on their behalf is unclear.[3] We can recognize in Martin's appointment, however, an early expression of an interest in western lands on the part of New Jersey men, which culminated, a few years later, in the purchase of a large tract in the southwest corner of present-day Ohio by Judge John Cleves Symmes.

For New York, William W. Morris. New York, and the three remaining states that sent federal surveyors to the West, in 1785, evinced no such interest in the Ohio country as did those states whose surveyors have been introduced, above. There is no

---

1783, with the rank of Lieutenant-Colonel. (Heitman, Register of Officers, p. 494.) For his election to surveyorship, see Jrnls. Cont. Cong., XXIX, 542. For his acceptance, see Papers Cont. Cong., LXXVIII, Pt. XXI, 405.

[1] Sherman to Governor of Connecticut, December 31, 1787, in New York Journal and Weekly Register, May 1, 1788; handwritten copy by C. A. Burton, Miscellaneous Collection, Western Reserve Historical Society, Cleveland, Ohio.

[2] Governor W. L. Livingston to Thomas Hutchins, Trenton, New Jersey, May 16, 1785, Hutchins Papers, Vol. III; and William Patterson, James Ewing et al., to Congress, Trenton, New Jersey, May 19, 1785, Papers Cont. Cong., XLII, Pt. V, 327. Martin had been a captain in the Continental Army. (Heitman, Register of Officers, p. 381.) For notice of his election to surveyorship, see Jrnls. Cont. Cong., XXVIII, 466.

[3] Although the list of signers who recommended Martin included men who are known to have been interested in investing in western lands, none of these men later appeared among the purchasers of land in the Seven Ranges. Over three hundred acres of land were purchased in Martin's own name, however. (See list of purchasers of land in the Seven Ranges, "Schedule of Sales of Lands in the Western Territory of the United States, at Public Auction, from the 21st September to the 9th October, 1787," Papers Cont. Cong., LIX, Pt. III, 135.) On a part of this land, a few miles upstream from a point opposite Wheeling, Martin founded the settlement known today as Martins Ferry, late in 1787.

evidence, for example, that New York's appointee, Lieutenant Morris, was expected to spy out the land.[1] Apparently an individual seeking an employment appropriate to his technical training, Morris was the single surveyor who could participate as an equal with Hutchins in "the Astronomical business of the Geographers Department." He gained Hutchins' special commendation for his assistance in 1785.[2]

For Virginia, Alexander Parker. Captain Parker apparently represented the Virginia county surveyor type--that class of woods-wise men to whose independent surveying activities Jefferson would have originally thrown open the entire public domain.[3] Declaring himself "sensible of the Honour done me" by the appointment as surveyor in the name of Virginia, Parker quite evidently viewed himself as a servant of the federal government, rather than the agent of a state interested in fostering colonization on federal lands.[4]

For Maryland, James Simpson. Maryland gave evidence of indifference toward the possibility of a first-hand report on federal lands by appointing an out-of-state surveyor, James Simpson of York County, Pennsylvania.[5] It may be conveniently noted here that, through Simpson, the party of surveyors encamped at the mouth of Little Beaver Creek made their only known contact with the boundary commissioners who established the beginning point for federal surveying. Simpson visited the Pennsylvania commissioners as

---

[1] Notice of Morris' election appears in Jrnls. Cont. Cong., XXXVIII, 398. He had left the Continental Army with the rank of Lieutenant. (Heitman, Register of Officers, p. 403.)

[2] Thomas Hutchins to Congress, New York, December 27, 1785, in Papers Cont. Cong., LX, 225. Morris, in later applying to succeed Hutchins as Geographer, wrote, "My studies have been directed principally to a knowledge of the Mathematics." (William Morris to President Washington, New York, May 12, 1789, Applications for Office under President Washington, Vol. XX, Papers of George Washington, Manuscripts Division, Library of Congress.)

[3] Parker, of Westmoreland County, Virginia, had served as Captain in the Continental Army. (Heitman, Register of Surveyors, p. 424.) For notice of election to surveyorship, see Jrnls. Cont. Cong., XXVIII, 398. While serving under Hutchins in 1785 he was noteworthy for his frontiersman's attitude and conduct. ("General Butler's Journal," Olden Time, II [October, 1847], 435; entry for September 30, 1785.)

[4] Papers Cont. Cong., LXXVIII, Pt. XVIII, 549.

[5] Simpson's place of residence is identified in entry for December 3, 1786, Journal of John Mathews.

they proceeded with the laying out of "Ellicott's Line," early in October, 1785.[1]

For Georgia, Robert Johnston. Dr. Johnston was another out-of-state appointee.[2] Georgia, a state rich in western lands at this time, had no reason to prefer the sending of a native son, for the sake of a report on distant federal lands. To send a surveyor at all was simply a cooperative gesture. Johnston himself--barring the possibility of mistaken identity--seems to have been a man of means who sought a surveyor's appointment for the purpose of selecting land for his own investment.[3]

By September 30, Hutchins, the eight surveyors identified above, and a retinue of about thirty helpers, were all assembled at the mouth of Little Beaver Creek. A visitor in their camp expected them to progress rapidly with their work, and yet he found cause for misgiving. Hutchins was openly apprehensive of Indian hostility, expressing himself as disposed to "instantly quit the business and return home" if danger threatened.[4] The initiative now lay with the Indians.

Surveying, 1785

Pursuant to his instructions in the Land Ordinance to personally attend to the running of the first east and west line, Hutchins, according to a witness, "made a beginning . . . at the post set up by Mr. Rittenhouse," September 30.[5] He proceeded westward until October 8, when, having surveyed less than four miles of this base line (Fig. 12), he suspended operations due to the receipt of "disagreeable intelligence" concerning the In-

---

[1] Thomas Hutchins to President of Congress, New York, November 24, 1785, Enclosure No. 1, Papers Cont. Cong., LX, 201-204.

[2] Johnston's place of residence was Baltimore, Maryland. (Robert Johnston to Congress, Pittsburgh, October 27, 1785, in Papers Cont. Cong., XLI, Pt. IV, 319.) For notice of his election to surveyorship, see Jrnls. Cont. Cong., XXVIII, 466.

[3] A "Doctr. Robt. Johnston" purchased about eighteen square miles of land in the Seven Ranges in 1787. ("Schedule of Sales of Lands in the Western Territory of the United States," Papers Cont. Cong., LIX, Pt. III, 137.) Hutchins uses the title "Doctor" in writing of Johnston in his Journal, Hutchins Papers, Vol. II.

[4] "Journal of General Butler," Olden Time, II (October, 1847), 435; entry for September 30, 1785.

[5] Ibid. This post, it will be recalled, had been established more than a month before this date.

dians.¹ Though the fact was not yet apparent, the season's surveying had come to an end: the surveyors had come west in vain.

The intelligence which reached Hutchins told of an Indian depredation at "Tuscarawas," the Delaware village located about fifty miles west of the beginning point of the surveys (Fig. 12).² At a trading post near the village, two traders had been set upon by a band of Indians, September 26. According to the report, stores of merchandise were plundered, one of the traders was killed, and "all the signs of War were left behind, [the Indians] having marked the inside of the door and many contiguous Trees with red Paint."³ Hearing this, Hutchins immediately retired from the line of survey, and supervised the shifting of the surveyors' camp to the south side of the Ohio River.

There was still hope, despite this news, that the Delaware and Wyandott, as parties to the Treaty of Fort McIntosh, would send chiefs to attend the surveyors, and thereby guarantee their safety. Hutchins had dispatched a messenger to these two tribes in September, but it was not until October 15 that a letter of response, "spoken by Captain Pipe for the Delawares and Wyandotts," was received.⁴ The letter, more apologetic than threatening in tone, simply declined Hutchins' invitation. Hutchins, who had said that he could not think himself and people safe without the chiefs, prepared almost at once to decamp.

It might seem strange that the troops at Fort McIntosh, who had been expected by Congress to protect the surveyors, were of no help on this occasion. This lack of support was due in part to the reduced strength of the garrison, of which the reader has been made aware. Unfortunately, the few remaining troops were needed at the site of a prospective treaty conference with the

---

¹Thomas Hutchins to President of Congress, New York, November 24, 1785, Papers Cont. Cong., LX, 193-200.

²This village, called the "foremost salient" of the Delaware, earlier in this study, lay at the southeast corner of the territory allowed the Indians under the Treaty of Fort McIntosh. Bolivar, Ohio, stands on the site of this village today.

³Hutchins to President of Congress, New York, November 24, 1785, Enclosure No. 1, Papers Cont. Cong., LX, 201-204.

⁴Hutchins had sent one William Wilson westward on the Sandusky Trail with an invitation in the second week of September. (Hutchins to President of Congress, Pittsburgh, September 15, 1785, Papers Cont. Cong., LX, 189-191.) Captain Pipe's answer appears in Hutchins to President of Congress, New York, November 24, 1785, Enclosure No. 2, ibid., pp. 209-212.

Shawnee, further down the Ohio River.[1] On the day before surveying began, all of the infantry based at Fort McIntosh had embarked for the treaty grounds, floating past the surveyors' camp early that evening.[2]

Deprived of army support, and disappointed by supposedly friendly Indian chiefs, Hutchins and his surveyors returned to Pittsburgh, where survey crews earlier recruited in that village were paid off and discharged. From Pittsburgh, the frustrated surveyors set forth on the Pennsylvania Road for their respective homes, with nothing but debts to show for their time and trouble.[3] Last to depart was Hutchins, who started off for New York November 1.[4]

Upon arriving in New York, Hutchins submitted to Congress a map showing the few miles of base line surveyed.[5] Perhaps in an attempt to give Congress a sense of value received for money expended, he added a copious description of the country traversed by the line.[6] The map has been lost, but the description survives, and from it the following excerpt has been taken:

> For the distance of Forty six Chains and Eighty six links West . . . , the Land is remarkably rich, with a deep black Mould, free from Stone, excepting a rising piece of ground on which there is an improvement of about 3 1/2 Acres, where there are a few Grey and Sand Stones thinly scattered. The whole of the above distance is shaded with black and white Walnut Trees, also with Black, Red and an abundance of white Oaks, some Cherry Tree, Elm Hoop-Ash, and great quantities of Hickory, Sassarfrax, Dogwood, and innumerable and uncommonly

---

[1] The treaty conference at the mouth of the Great Miami River led to the signing of the Treaty of Fort Finney, January 31, 1786. The Delaware and Wyandott had learned of plans for this conference a few days before Hutchins' messenger arrived, and it was this news which Captain Pipe used as an excuse for declining Hutchins' invitation. (Ibid.)

[2] "Journal of General Butler," Olden Time, II (October, 1847), 434; entry for September 29, 1785.

[3] The comings and goings of the surveyors, as well as their financial transactions, are made apparent in the accounts which they submitted to Congress late in 1785. See Papers Cont. Cong., XLI, Pt. IV, 305, 307, 309, 315, 317.

[4] Journal of Thomas Hutchins, entry for November 1, 1785.

[5] Hutchins to President of Congress, New York, December 27, 1785, Papers Cont. Cong., LX, 225.

[6] "A brief account of the Soil and Timber in that part of the Western Territory through which an East and West Line has been surveyed," ibid., pp. 229-236.

large Grape Vines producing well tasted Grapes of which Wine may be made. All the Hills in this part of the Country seem to be properly disposed for the growth of the Vine. Near the termination of the above mentioned measurement is a thicket of Shoemack, Hazel and Spice bushes, through which a passage was cut for the Chain-carriers. The first of these Bushes produces an Acid berry well answering the purposes of sowering for Punch, the Hazel yield an abundance of Nuts, and the Spice bushes bear a berry, red when ripe of an aromatic smell, as is also the Shrub on which it growes; the berry is about the size of a large Pea, of an Oval shape possessing some Medicinal virtues, and has often been used as a substitute for Tea by sick and indisposed persons. The Dogwood, the bark is used by the inhabitants and is said to be little inferior to Jesuits bark in the cure of Agues; the Tree produces a berry about the size of a large Cramberry when ripe, but something longer and smaller toward the Ends, excellent for bitters; and decoctions made of the budds or blossoms have proved very salutary in several disorders, particularly in Bilious complaints.

The whole of the above described Land is too rich to produce Wheat, the aforementioned rising ground excepted, but it is well adapted for Indian Corn, Tobacco, Hemp, Flax, Oats &c and every species of Garden Vegetables, it abounds with great quantities of Pea Vine, Grass, and nutritious Weeds of which Cattle are very fond, and on which they soon grow fat.[1]

Hutchins covered eight closely-written foolscap pages with observations of this kind, thereby setting an example of verbosity which may never have been equalled in the subsequent history of the public land surveys.

Despite the dishearteningly inconsequential nature of the first attempt at surveying under the Land Ordinance, Congress had not yet lost faith in the enterprise. In May, 1786, Hutchins was given, in effect, a vote of confidence, when Congress passed a resolution authorizing the Geographer to try again.[2]

Surveying, 1786

Prospects for surveying in 1786 were brightened by the construction of a new fort, Fort Harmar, on the Ohio River at the mouth of the Muskingum (Fig. 12), and by the success of efforts directed toward increasing the strength of the Army.[3] Further,

---

[1] Ibid., pp. 229-230.

[2] Resolution of May 9, 1786, Jrnls. Cont. Cong., XXX, 248.

[3] Fort Harmar had been established with a view to enforcing the removal of "squatters" from national lands, as well as for the protection of the surveyors. (Harmar to Knox, Philadelphia, October 22, 1785, in Butterfield, Journal of Jonathan Heart, pp. 92-94.) For a description of the Army's condition during this period in its history, see James Ripley Jacobs, The Beginning of the U.S.

the treaty conference with the Shawnee had been brought to a seemingly successful conclusion, and the Wyandott and Delaware, who attended the conference, appeared to be resigned to the survey of their ceded lands.[1] The trustworthiness of these favorable indications remained to be tested.

Renewed preparations. The Geographer returned to Pittsburgh, in June, 1786, expecting to be able to report to Congress, at the end of a season's work, that thirteen ranges of townships had been completed.[2] He seemed justified in this hope for several reasons. First, Congress had curtailed the north-south extent of the ranges. The resolution which authorized resumption of the surveys ordered that surveying be confined to the area south of the east-west line which Hutchins had begun to lay out in 1785.[3] Second, the prospective work involved in surveying had been simplified by the repeal of one of the important clauses which the Land Ordinance owed to Thomas Jefferson; upon a motion by Rufus King of Massachusetts, Congress had resolved to repeal the requirement that boundaries be run "by the true meridian."[4] Third, Hutchins succeeded almost at once in dispatching an invitation to chiefs of the Delaware and Wyandott, nearly three months in advance of the schedule of the preceding year. Fourth, Hutchins had been led to expect that a full complement of thirteen surveyors,

---

Army, 1783-1812 (Princeton, New Jersey: Princeton University Press, 1947), pp. 13-39.

[1] After the signing of the Treaty of Fort Finney, January 31, 1785, General Butler, one of the American commissioners, took leave of the Indians, convinced of their "perfect satisfaction" with the terms agreed upon. ("General Butler's Journal," Olden Time, II [December, 1847], 531; entry for February 4, 1786.)

[2] Hutchins to President of Congress, Camp at the Intersection of the West Bounds of Pennsylvania with the River Ohio, August 13, 1786, Papers Cont. Cong., LX, 249-252.

[3] Resolution of May 9, 1786, in Jrnls. Cont. Cong., XXX, 248. This limitation on the extent of federal surveying was almost certainly instituted in anticipation of surveys by the State of Connecticut. Confirmation of Connecticut's title to the Western Reserve, north of the forty-first parallel (Fig. 14), was under debate at this time. For approval of Connecticut's claim, May 26, 1786, see ibid., p. 310.

[4] Resolution of May 12, 1786, in Jrnls. Cont. Cong., XXX, 252. In surmising the background for this repeal, one should not overlook the fact that Thomas Hutchins had been in close touch with members of Congress earlier in the same year. (Hutchins to George Morgan, New York, February 7, 1786, George Morgan Papers, Manuscripts Division, Library of Congress.)

one to be assigned to each range of townships, would appear for service in this year of renewed effort.[1]

As the surveyors arrived in Pittsburgh, they once more undertook "purchasing provisions, hireing Men and Horses, &c.," and this time they extended their recruiting and procurement activities down the Ohio, among the pioneer settlements on the Virginia shore.[2] When preparations were complete, each of the states but Delaware was represented by a surveyor equipped and ready to take to the field.[3] Four states--Rhode Island, Pennsylvania, North Carolina and South Carolina--were represented for the first time, and two states--New Hampshire and Virginia--were now served by new men. Of the six men thus added to the roster of pioneer federal surveyors, four deserve special note:[4]

(1) Winthrop Sargent, surveyor for New Hampshire and successor to Edward Dowse. Major Sargent, whose application for a surveyorship was sponsored by the Secretary of War, was a Massachusetts man soon to be elected Secretary of the newly organized Ohio Company of Associates. Seizing an opportunity for exploration offered by a delay in getting the surveys under way, Sargent reconnoitred the lower valley of the Muskingum River--the heart of the district soon to be purchased by the Ohio Company.[5]

(2) Ebenezer Sproat, surveyor for Rhode Island. Colonel Sproat was soon to be identified with the Ohio Company. As a representative of Rhode Island, he later received an appointment as surveyor of Ohio Company lands, and was counted

---

[1] Hutchins to President of Congress, August 13, 1786, Papers Cont. Cong., LX, 249.

[2] Diary of Winthrop Sargent, entry for July 21, 1786.

[3] Journal of Thomas Hutchins, entries for July 5-July 21, 1786.

[4] The other two men were Charles Smith, who took the place of Alexander Parker of Virginia, and Samuel Montgomery, a man already present on the frontier, who was sworn in as surveyor for North Carolina in the place of the absent Absolom Tatom. Six men who had come west in 1785 returned: Benjamin Tupper (Massachusetts), Isaac Sherman (Connecticut), William Morris (New York), Absalom Martin (New Jersey), James Simpson (Maryland) and Robert Johnston (Georgia).

[5] Henry Knox to Hutchins, New York, June 4, 1786, Society Miscellaneous Collection, Historical Society of Pennsylvania, Philadelphia; Sargent to Samuel Parsons, Fort Harmar, August 1, 1786, Samuel Parsons Papers, Western Reserve Historical Society, Cleveland. For Major Sargent's record in the Continental Army, see Heitman, Register of Officers, p. 481. Sargent's tour of duty as a surveyor has been covered in Benjamin H. Pershing, "A Surveyor in the Seven Ranges," Ohio State Archaeological and Historical Quarterly, XLVI (1937), 257-270.

among the company of men who landed at the mouth of the Muskingum River to found Marietta, in 1788.[1]

(3) Adam Hoops, surveyor for Pennsylvania. Colonel Hoops, a professional surveyor and land speculator from Philadelphia, was a friend of Hutchins', and probably owed his appointment to Hutchins' influence.[2]

(4) Israel Ludlow, surveyor for South Carolina. Ludlow, appointed to fill a vacant surveyorship, was a young man from New Jersey who came west in 1786 to make his fortune on the frontier. Whatever his connection may have been with speculators from his home state at this time, he later became actively interested in the Symmes Purchase. Before his death, in the early 1800's, Ludlow surveyed more land in the Ohio country than any other federal surveyor.[3]

Once the Geographer and surveyors were assembled at the intersection of Pennsylvania's western boundary and the Ohio River, surveying would have begun immediately but for lack of security against Indian attack. The Indians delayed their response to Hutchins' invitation, and Colonel Harmar, still the general commanding officer in the West, proved reluctant to assign an armed escort to the surveyors out of fear of provoking an Indian war. A letter from the Secretary of War, and a flat refusal on the part of the surveyors to proceed without protection, however, combined to overcome Harmar's caution, and parts of three companies of infantry, totalling about one hundred and fifty men, were placed at Hutchins' disposal.[4] On the day these

---

[1] Colonel Sproat replaced Caleb Harris, who had failed to come west both in 1785 and in 1786. For notice of Sproat's appointment as an Ohio Company surveyor, November 23, 1787, see Archer B. Hulbert (ed.), The Records of the Proceedings of the Ohio Company ("Marietta College Historical Collections, Ohio Company Series," Vol. I; Marietta, Ohio: Marietta Historical Commission, 1917), p. 26. For his record in the Continental Army, see Heitman, Register of Officers, p. 513.

[2] Hutchins had earlier made an employment request on behalf of Hoops. (Hutchins to John Montgomery, Philadelphia, May 26, 1784, John Montgomery Papers, Chicago Historical Society, Chicago.) Hoops-Hutchins correspondence may be found in Hutchins Papers, Vol. III. For Hoops' record in the Continental Army, see Heitman, Register of Officers, p. 300.

[3] Ludlow replaced William Tate, who had failed to come west both in 1785 and in 1786. From Morris County, New Jersey, Ludlow went west as a practiced surveyor, expecting to stay. (Robert Morris to Timothy Pickering, New Brunswick, New Jersey, July 18, 1796, and Jonathan Dayton to Pickering, Elizabethtown, New Jersey, July 18, 1796, Applications for Office under President Washington, Vol. XVIII.) Ludlow is the subject of repeated reference in subsequent chapters of the present study.

[4] Hutchins to President of Congress, Camp at the Intersec-

troops arrived, August 5, Hutchins and a few of the surveyors crossed the Ohio River to the federal shore, from their encampment opposite the beginning point.[1]

Surveying resumed. Surveying began again, August 9, 1786. From that date to September 18, Hutchins pushed steadily westward, marking a course which approximated a parallel of latitude. He reached a point six miles from the Pennsylvania boundary on the second day, and here Absalom Martin of New Jersey directed a line southward, setting out independently to complete the first township.[2] At the end of each succeeding interval of six miles, other surveyors started south, in an order determined by lot. According to plan, each of the surveyors thus launched was to be responsible for an entire range of townships.[3]

By the end of August, Hoops, Sherman and Sproat had followed Martin's example.[4] Then Sargent and Simpson took their turns, and Morris was about to set off on his assigned strip of country--the Seventh Range--when the first sign of trouble appeared.[5] On September 13, a message reached Hutchins, in which the chiefs of the Delaware and Wyandott declined, for the second and last time, to come forward and guarantee the safety of the surveyors.[6] Hutchins continued westward none the less, dispatching Morris to work on his appointed range three days later. Hutchins now advanced into the Eighth Range, transferring his camp ahead to a convenient creek, which not only brought his party into the immediate neighborhood of "Tuscarawas," but separated them by about forty-five miles from their principal military support (Fig. 12). Despite the fact that three companies of infantry

---

tion of the West Bounds of Pennsylvania with the River Ohio, August 13, 1786, Papers Cont. Cong., LX, 249-252.

[1] Journal of Thomas Hutchins, entry for August 5, 1786.

[2] Ibid., entry for August 11, 1786.

[3] Diary of Winthrop Sargent, entry for July 14, 1786.

[4] Journal of Thomas Hutchins, entry for September 2, 1786.

[5] Ibid., entries for September 6 and 11, 1786; Diary of Winthrop Sargent, September 9, 1786; and Hutchins to President of Congress, Camp at the junction of Wheeling Rivulet and the Ohio, October 12, 1786, Papers Cont. Cong., LX, 261-271.

[6] Jacob Springer to Hutchins, 38 Miles on the East and West Line, September 13, 1786, Papers Cont. Cong., LX, 257-260.

had been assigned to Hutchins, all but thirty soldiers, under the command of a lieutenant, were confined to a camp on the Ohio River, for want of supplies.[1] The Geographer and his followers were now in a dangerously exposed position.

On the morning of September 18, the Geographer's camp awoke to find that a pole marking the conclusion of the previous day's surveying had been broken during the night, apparently as a warning from hostile natives.[2] That afternoon, an express from the Army camp on the Ohio brought intelligence that warriors were gathering at the Shawnee towns, about one hundred and fifty miles to the southwest, intending "to cut Hutchins off and all his Men."[3] Thoroughly alarmed, Hutchins prepared to retreat, and sent messages to the surveyors on their several ranges, asking them to lose no time in following his example.[4]

Retreat and reorganization. The retreat which followed was almost comic in its confusion. Hoops, on the Second Range, had already retired from the field, due to ill health.[5] Sherman and Sproat, on the Third and Fourth Ranges, failed to learn of the Indian alarm, and returned to the Ohio River simply out of a need to replenish their supplies.[6] Sargent, on the Fifth Range, hearing that "the Geographer had run away and all the surveyors after him," viewed the proceedings with scorn, and was persuaded only with difficulty to leave his work.[7] At length, however, the surveyors were collected together at the house of William McMahon, on the Virginia shore, and all of the troops were concentrated at a fortified position on the federal side of the Ohio, downstream from the point of beginning.[8]

---

[1] Hutchins to President of Congress, Camp at the junction of Wheeling Rivulet and the Ohio, October 12, 1786, Papers Cont. Cong., LX, 261-271.

[2] Journal of Thomas Hutchins, entry for September 18, 1786.

[3] Depositions of George Brickell and Thomas Girty, Papers Cont. Cong., LX, 277-278.

[4] Hutchins to President of Congress, October 12, 1786, Papers Cont. Cong., LX, 262.

[5] Journal of John Mathews, entry for September 1, 1786.

[6] Diary of Winthrop Sargent, entry for October 1, 1786.

[7] Ibid., entries for September 20-24, 1786.

[8] Ibid., entries for September 27-October 3, 1786; and Hutchins to President of Congress, October 12, 1786, Papers Cont. Cong., LX, 261-271.

By the beginning of October, the settled objective for the season had become the completion of four ranges of townships. The question of greatest importance now was whether the surveyors, in taking to the woods to accomplish this limited objective, would simply risk harassment by renegade bands of Indians, or would face an attack by the large war party which was reportedly being organized in the Shawnee towns.[1] That the scheduled attack by this body of warriors never occurred was almost certainly due to the fact that, early in October, an expedition of Kentucky militia under the command of Colonel Benjamin Logan suddenly descended upon the Shawnee country, spreading terror and destruction.[2] The Kentuckians, in executing an act of local vengeance, apparently permitted the completion of the first effective season of national surveying.

With all of the troops at his disposal sufficiently provisioned to take to the field, for the first time, Hutchins arranged for the resumption of surveying, apparently without knowledge of Colonel Logan's impending attack upon the Shawnee.[3] The First Range having been surveyed in September, Hutchins sent six surveyors into the succeeding three ranges whose completion was now expected, and as a concession to the headstrong Sargent he allowed that surveyor to venture into the Fifth Range once more.[4] About eighty soldiers were distributed among the surveyors for their immediate protection, and the remainder were held in reserve in a central position (Fig. 12), behind hastily erected earthworks.[5]

Four ranges completed. By the middle of November, four ranges of townships were finished, without Indian incident. On the Fifth Range, however, Sargent's work was cut short by the stealth of a small band of Indian marauders. During a storm one

---

[1] Intelligence report of Captain Ferguson, Fort Pitt, September 14, 1786, ibid., pp. 279-280.

[2] Colonel Harmar to the Secretary of War, Fort Hamar, November 15, 1786, in William Henry Smith, The Life and Public Services of Arthur St. Clair with His Correspondence and Other Papers (2 vols.; Cincinnati: Robert Clarke and Co., 1882), II, 19-20.

[3] Journal of Thomas Hutchins, entry for October 1, 1786.

[4] Ibid., entries for October 3-11, 1786; Diary of Winthrop Sargent, entries for October 5-11, 1786.

[5] Journal of Thomas Hutchins, entries for October 10, 15 and 16, 1786.

night, all of the horses accompanying Sargent's party of thirty-six men were driven off, except "one meagre thing" belonging to Sargent himself. Failing to secure replacements, Sargent abandoned the field, leaving his range in a half-finished condition.[1]

With this much accomplished, Hutchins seriously considered rounding out all of the seven ranges upon which work had been begun, but the surveyors were generally averse to the idea, nighttime temperatures now having dropped to the freezing point.[2] The troops, many of them "barefoot and miserably off for clothing," were in no condition to continue on field duty.[3] In consequence, the soldiers were allowed to embark for winter quarters at Fort Harmar, and the surveyors retired to the comfort and security of McMahon's house, on the Virginia shore.[4]

Paper work. At McMahon's house, Hutchins soon set about marshalling the documentary evidence of the surveys. Under his direction, the notes which the surveyors had taken in the field, for the first four ranges, were transcribed and rearranged in a form suitable for submission to the Board of Treasury.[5] Martin, Sherman and Sproat, to whom had fallen the official responsibility for surveying the first four ranges, stayed at McMahon's (Fig. 12) until their signatures could be affixed to the completed transcriptions.[6]

---

[1] Ibid., entry for October 22, 1786; Diary of Winthrop Sargent, entries for October 27-November 6, 1786; Colonel Harmar to Hutchins, Fort Harmar, November 6, 1786, George Morgan Papers, Manuscripts Division, Library of Congress.

[2] Journal of Thomas Hutchins, entries for November 10 and 11, 1786; Diary of Winthrop Sargent, entries for October 23 and November 1, 1786; Hutchins to President of Congress, Ohio County, Virginia, December 2, 1786, Papers Cont. Cong., LX, 281-283.

[3] Colonel Harmar to the Secretary of War, Fort Harmar, November 15, 1786, in Smith, St. Clair Papers, II, 19-20.

[4] Entry for November 25, 1786, Journal of Joseph Buell, in Hildreth, Pioneer History, p. 148; entries for November 8, 14, 15 and December 3, 1786, Journal of John Mathews. For threatening message from the Delaware and Wyandott, which further encouraged abandonment of the field, see Indian chiefs to General Butler, Sandusky, October 28, 1786, Papers Cont. Cong., LX, 281-282.

[5] Survey notes for the first four ranges of townships in the Seven Ranges may be found in four calf-bound volumes, in Records of the General Land Office (Record Group 49), Cartographic Records Branch, the National Archives. The notes have been so transcribed that the data for the four sides of any one township may be read in sequence. The townships follow an order of numbering based on the Ohio River.

[6] Although the signatures of only these three men appear on

By the first week in December, nearly all of the surveyors had departed for their homes in the East, and the final stage of operations had begun. Hutchins, writing to the President of Congress that his departure from the Western Territory would be delayed "until such time as the Townships are delineated on paper," was referring to the preparation of plats--the drawings of the boundaries of each township which the Land Ordinance required.[1] These remained to be drafted after the surveyors' notes had been set in order. It is believed by the present author that the plats, save for those representing townships in the Third Range, were drawn in the course of the following two months either by Hutchins or under his direct control.[2] Absalom Sherman, who withdrew to the house of Charles Wells (Fig. 12), ten miles distant from McMahon's, is believed to have prepared the plats for the Third Range.[3]

On January 27, 1787, seven months after his arrival in the West for a second attempt at surveying, Hutchins departed from his quarters on the Ohio River.[4] Leaving Pittsburgh four days later, he arrived in New York February 21, ready to present to the Board

---

the transcribed survey notes, the field work of three additional men, Tupper, Hoops and Simpson, is represented. (Sherman to Hutchins, Ohio County, Virginia, Hutchins Papers, Vol. III.) Further, the handwriting of at least two men in addition to that of the signatory surveyors appears in the notes. Evidence of John Mathews' participation in the preparation of the notes appears in entries, November 15-December 3, 1786, Journal of John Mathews. For a record of the departure dates of the several surveyors for their homes in the East, see ibid.

[1] Hutchins to President of Congress, December 2, 1786, Papers Cont. Cong., pp. 281-283.

[2] This judgment is based upon an examination of the plats themselves, in Records of the General Land Office (Record Group 49, Cartographic Records Branch, the National Archives. The plats for townships of the First, Second and Fourth Ranges are all strikingly similar, and in definite contrast to those for the Third Range. They bear a distinct resemblance to a map entitled, "A Plan of the Several Villages in the Illinois Country," executed by Hutchins, which appears in Hutchins, A Topographical Description (1904), facing p. 112.

[3] The plats for the Third Range are distinguished by style of compass rose and graphic scale, manner of shading, form of lettering and other details. Sherman's removal to the house of Charles Wells is noted in Journal of Thomas Hutchins, November 21, 1786. His continued presence at Wells' house is noted in Journal of John Mathews, December 26, 1786.

[4] Ibid., entry for January 27, 1787.

of Treasury "the Plats and descriptions of four Ranges completely surveyed into Townships containing in the whole six hundred and seventy five thousand four hundred and eighty Acres."[1] Hutchins later declared, in a locution characteristic of the period, that he "flattered himself that he had performed his duties to the entire satisfaction of Congress."[2]

Surveying, 1787-1788

Hutchins may have vindicated his own actions to the satisfaction of Congress, but that body had understandably lost faith in the public land surveying system, by the spring of 1787. With only four ranges of townships ready to be advertised for sale after a lapse of nearly two years, Congress was prepared to consider the sale of large tracts of land without prior survey as a means of realizing an immediate income from the national domain.[3] Finding that Congress now expected no more of the federal surveyors than the completion of the Seven Ranges, Hutchins applied for leave to fulfill an engagement elsewhere.[4] His request was granted, and the task of completion was left to such of the surveyors of the preceding year as might be willing to assume the risks involved in the venture.[5]

Events in the field. First in the field, in 1787, were Absalom Martin and Israel Ludlow, who had wintered on the Ohio River.[6] They went into the woods early in April, and were followed

---

[1] Hutchins to President of Congress, New York, February 22, 1787, Papers Cont. Cong., LX, 293. Hutchins later submitted a map, now lost, covering the four ranges of townships. (Hutchins to President of Congress, April 18, 1787, Papers Cont. Cong., LX, 301.)

[2] Hutchins to President of Congress, March 19, 1787, Papers Cont. Cong., LX, 297.

[3] On the sale of land in the first four ranges see chap. ix, and on the sale of land in large tracts beyond the Seven Ranges, chap. x, below.

[4] Hutchins to President of Congress, New York, June 25, 1787, Papers Cont. Cong., LX, 185. The assignment to which Hutchins turned was the survey of a boundary within New York State: a meridian, westward of which New York retained jurisdiction while Massachusetts held land-title.

[5] For note on the release of Hutchins from public land surveying, see Jrnls. Cont. Cong., XXXII, 308, n. Among the surveyors who saw little point in returning to work in the Seven Ranges under existing conditions was Adam Hoops of Philadelphia. (Hoops to Hutchins, Philadelphia, May 7, 1787, Hutchins Papers, Vol. III.)

[6] Martin, it will be recalled, represented New Jersey, and

within two weeks by James Simpson, who had returned to the West from his home in York County, Pennsylvania.[1] Two other surveyors later appeared on the scene, but these three men appear to have pre-empted the surveying which remained to be done.[2] Throwing caution to the winds, they led their survey parties into the interior without an armed escort. They encountered no organized resistance, despite a formal notice which had recently been sent to Congress by the tribes of the Ohio country, asking that further surveying be prohibited.[3] Ludlow, however, lost his horses, as had Sargent in 1786, to a party of Indian thieves, and all three surveyors retired to the Ohio River upon receiving reports that marauders were at large to the south of their immediate field of operations.[4] By the middle of May, they were applying for the protection of the Army.

The surveyors expected aid from a new army post which seemed to be ideally suited to their purposes. This was Fort Steuben (Fig. 12), located on the Ohio River within the First Range, to which a force of ninety men had been assigned during the previous winter.[5] Colonel Harmar, however, was holding this

---

Ludlow represented South Carolina. Ludlow had already distinguished himself from his fellows by being the only surveyor willing to venture as far west as the Seventh Range, in November, 1786. (Journal of Thomas Hutchins, entries for November 10 and 11, 1786.)

[1] Entries for April 10 and 21, 1787, Journal of John Mathews, in Samuel P. Hildreth, <u>Pioneer History: Being an Account of the First Examinations of the Ohio Valley and the Early Settlement of the Northwest Territory</u> (Cincinnati: H. W. Derby and Co., 1848), p. 178. Hildreth, in reproducing extracts from the Mathews Journal (ibid., pp. 170-192), goes well beyond the final date of the manuscript Journal in the Marietta College Library, heretofore cited. Henceforward, citations of entries after April 21, 1787, will refer to Hildreth's reproduction, as will be indicated.

[2] The two extra surveyors were Charles Smith of Virginia and Isaac Sherman of Connecticut. (<u>Ibid.</u>, pp. 179-180; entries for April 17 and May 15, 1787.)

[3] Speech of the United Indian Nations, at their Confederate Council, held near the mouth of the Detroit River, November 28 and December 18, 1786, in <u>American State Papers</u>, <u>Indian Affairs</u>, I, 8-9.

[4] Colonel Harmar to Secretary of War, Fort Harmar, May 14, 1787, Papers of General Josiah Harmar, William L. Clements Library, Ann Arbor, Michigan, Vol. XXVIII. Hereafter referred to as Harmar Papers.

[5] Major Hamtramck to Colonel Harmar, Fort Steuben, May 22, 1787, Harmar Papers, Vol. V.

detachment in readiness for removal to Vincennes.[1] As a sign of the times, Harmar was more interested in extending American influence further down the Ohio than in accommodating the federal surveyors, but he responded to their request by sending up sixty men from Fort Harmar.[2] After making a rendezvous at a point opposite Wheeling, these troops set off with the surveyors to cover them in the completion of their work.[3]

Within two weeks after resuming operations, Israel Ludlow finished the Seventh Range, striking the Ohio River about seven miles above the mouth of the Muskingum, where Fort Harmar was located.[4] Simpson and Martin brought the Sixth and Fifth Ranges, respectively, to completion soon after, and the escorting troops were able to rejoin their companies at Fort Harmar before July 10.[5] Although incidents of scalping and horse-thieving occurred in their vicinity both during and after this period, the surveyors were untroubled in their final efforts by Indian marauders.[6]

Paper work. Once again, William McMahon's house on the Ohio was the scene of record preparation. Martin, Ludlow and Simpson stayed here until the end of August, rearranging and transcribing their notes.[7] Ludlow may have been able to complete all of his paper work--both notes and plats--for the Seventh Range, during this time, but Martin and Simpson were prevented from completing their notes for the Fifth and Sixth Ranges by the lack of field notes for surveying done in 1786.[8] After their so-

---

[1] Harmar to Secretary of War, Fort Harmar, May 14, 1787, Harmar Papers, Vol. XXVIII.

[2] Surveyors to Harmar, Mr. McMahon's, Ohio County, Virginia, May 25, 1787, and Harmar to Surveyors, Fort Harmar, June 2, 1787, Harmar Papers, Vol. XXVIII.

[3] Entries for June 6 and 8, 1787, Journal of John Mathews, in Hildreth, Pioneer History, p. 181.

[4] Major Doughty to Colonel Harmar, Fort Harmar, June 24, 1787, Harmar Papers, Vol. VI.

[5] Major Doughty to Colonel Harmar, Fort Harmar, July 10, 1787, Harmar Papers, Vol. VI.

[6] Entries for June 23 and August 4, 1787, Journal of John Mathews, in Hildreth, Pioneer History, pp. 182-183.

[7] Ibid., pp. 182, 186; entries for July 31 and September 3, 1787.

[8] The notes and plats for Ranges Five, Six and Seven may be found with the records for the first four ranges of the Seven Ranges, in Records of the General Land Office (Record Group 49),

journ at McMahon's, the three surveyors proceeded to New York, where the content of these earlier notes was apparently incorporated by Hutchins into the records for 1787.[1]

By the middle of September, 1787, Hutchins, the three final surveyors, and the records of survey, were all assembled in New York, yet the Board of Treasury, also located in New York, did not receive the finished notes and plats until July of the following year.[2] Martin was probably mainly responsible for the delay, since he, piqued by the Board of Treasury's refusal to allow him an extra allowance for "protracting the townships," carried his plats back to the Ohio River, there to complete them at leisure.[3] For his part, Hutchins seems to have been primarily interested in making out a report on the surveying assignment which had occupied him during the summer of 1787.[4] After Martin's return from the West, Hutchins further delayed submission of notes and plats for the final three ranges until he had prepared a general plan covering all of the Seven Ranges. At last, July 26, 1788, Hutchins submitted the general plan and the concluding notes and plats to the Board of Treasury, and the first phase of U.S. public land surveying came to an end.[5]

Cartographic Records Branch, the National Archives. Both the notes and plats for Range Seven appear to the present author to be the work of Israel Ludlow. Ludlow's hand appears with those of Martin and Simpson in the notes for Range Six, and with those of Martin and Hutchins in the notes for Range Five. The township plats for Ranges Five and Six appear to be, with two exceptions, the work of Simpson and Martin, in keeping with the signatures thereon.

[1] This judgment is supported by the occurrence of Hutchins' handwriting in the notes for Ranges Five and Six.

[2] Evidence for the presence of Hutchins and the surveyors in New York may be found in Memorial of Surveyors to Congress, New York, September 22, 1787, Papers Cont. Cong., XLI, Pt. IX, 461, and Hutchins to President of Congress, September 24, 1787, Papers Cont. Cong., LX, 323.

[3] Martin to Hutchins, New York, October 3, 1787, Hutchins Papers, Vol. III; and entry for November 30, 1786, Journal of John Mathews.

[4] Hutchins submitted final returns on the meridian in New York which he had helped survey, early in 1788. (Hutchins to President of Congress, February 4, 1788, Papers Cont. Cong., LX, 327.)

[5] Hutchins to Commissioners of the Board of Treasury, New York, July 26, 1788, Records of Thomas Hutchins, Records of the General Land Office, Interior Section, Natural Resources Records Branch, the National Archives. On deposit with this letter is an account of the acreage of Ranges Five, Six, and Seven, in Hutchins' hand.

CHAPTER VII

THE QUALITY AND COST OF THE FIRST SURVEYS

Deserving of special attention are two subjects which were omitted from the story which has just been told: the quality and the cost of surveying. The quality of the work performed in the field, consisting simply of the accuracy with which the surveyed lines were laid down, will be discussed in the opening pages of the present chapter. The remaining pages will be concerned with the expenses which a reluctant Congress paid, in support of the activities of the Geographer and surveyors.

The Quality of Surveying

The Geographer's Line. The foundation upon which the township boundaries in the Seven Ranges were constructed was the line which Hutchins initiated in 1785, and ran westward in 1786 until caused to flee the field. Called simply the East and West Line, at the time of survey, it has come to be known as the Geographer's Line, in honor of Hutchins (Fig. 12). In constructing it, Hutchins was called upon to meet two challenges to his capacity for accurate surveying. He was required, first, to determine the latitude of the point of beginning, and second, to make his line conform to a parallel of latitude.[1]

At the beginning of the Geographer's line, Hutchins reported that he "made the Latitude 40° 38' 02" from a mean of a great number of observations both on the Sun and North Star."[2] Since the beginning point lay in the course of a meridian already

---

[1] In the language of the Land Ordinance, the Geographer was to "take the latitude of the extremes of the first north and south line," and "attend to the running of the first east and west line." (Carter, Territorial Papers, II, 13.)

[2] Hutchins to President of Congress, New York, November 24, 1785, Papers Cont. Cong., LX, 194. Despite his instructions in the Land Ordinance, Hutchins did not attempt to take the latitude of the other extreme of "the first north and south line," that is, the northern end of Pennsylvania's western boundary. This was done by Pennsylvania's boundary commissioners in 1876.

marked by the Pennsylvania boundary commissioners, it may be assumed that Hutchins availed himself of its convenience, and confined his observations of both the Sun and the North Star to their altitudes at the times of their crossing this meridian.[1] Despite the number of observations made, Hutchins was necessarily limited, in point of accuracy, by the capabilities of his instrument. That he erred nearly half a mile in locating his position suggests that he employed a sextant, an instrument in common use at that time.[2]

The second problem, that of laying down a parallel of latitude, was familiar to Hutchins if only because he had assisted in extending Pennsylvania's southern boundary westward, in 1784.[3] Along this state boundary, as in modern practice, points on the curved parallel of latitude were located on the ground by measuring carefully calculated distances toward the north or south from straight, surveyed lines (Fig. 13A).[4] Probably, however, Hutchins ignored this precedent in the first season of public land surveying, contenting himself with approximating a curved line by laying down a series of short compass courses.[5] In the second season of

---

[1] The Pennsylvania boundary commissioners themselves had determined the latitude of various points along the meridian by measuring the altitude of Polaris and of selected stars in the southern sky at the instant of their crossing the local meridian in the sky. (Entries for July 1 and 10, 1785, and July 8, 1786, Journal of Andrew Porter, in Porter, "Life of Porter," *Pennsylvania Magazine of History and Biography*, IV [1880], 272, 273, 281.)

[2] Hutchins' error amounted to 25.2 seconds of arc, or about 2,570 feet. The idea of connecting this magnitude of error with the use of a sextant is based upon an observation of Jared Mansfield on sextants in Mansfield to Secretary of the Treasury, Cincinnati, September 13, 1806, in Carter, *Territorial Papers*, Vol. VII: *Territory of Indiana, 1800-1810* (1939), p. 391.

[3] Hutchins' participation in this boundary survey has been mentioned in note under "The Place of Beginning," chap. iv, above.

[4] As shown in Fig. 13A, stations on a curved parallel of latitude may be established by offsets from a connected series of straight lines. The kind of series shown here characterizes the "secant method," which Hutchins and his colleagues used in extending the Mason and Dixon Line westward in 1784, according to Report to President of the State of Pennsylvania, Philadelphia, December 23, 1784, Hutchins Papers, Vol. III.

[5] Reliance upon the compass needle is implied by Hutchins' care in reporting the variation of his compass from true north, and by his failure to note a latitude reading anywhere save at the initial point of the Geographer's Line, in his description of that line, Papers Cont. Cong., LX, 236.

surveying he may have taken more care. The state-appointed surveyors who in this second year laid out segments of the line under Hutchins' supervision, wrote of the "great accuracy" required.[1] Still, their survey notes fail to refer to offsets of the kind made along Pennsylvania's southern boundary. Even if we assume that evidence of such offsets was for some unknown reason removed, in the course of transcribing the official notes, the fact remains that the Geographer's Line ended about fifteen hundred feet south of the latitude of its beginning point.[2] The magnitude of the error suggests, again, that Hutchins employed a sextant in checking his latitude as he advanced westward.[3]

In Jefferson's scheme for western states, there would have been many equivalents of the Geographer's Line, the first of them supposedly cutting through the midst of the "State of Washington," as Hutchins' line did, and the others serving to bound additional states. In all cases, they would have functioned as base lines for the public land surveys. It is impossible to say, of course, whether the state boundaries of this unrealized scheme would have been run after the example of Pennsylvania's southern boundary, or in the less accurate manner that characterized Hutchins' work.

Township boundaries. In laying out township boundaries south of the Geographer's Line, each of the state-appointed surveyors had to face the challenge of consistently measuring both directions and distances. None of the surveyors acquitted himself well in executing either kind of measurement.

In directing their lines of sight, the surveyors relied upon the circumferentor.[4] Relieved by Congress of the necessity of running township boundaries "by the true meridian," they apparently met the problem of orientation in the simplest possible

---

[1] Memorial of Surveyors to Congress, Banks of the Ohio, August 14, 1786, Papers Cont. Cong., XLI, Pt. IX, 415-416.

[2] For position of beginning and end of Geographer's Line, compare the following two topographic sheets of the U.S. Geological Survey: Dover, Ohio, Quadrangle (1912 ed.), and Wellsville, Ohio, Quadrangle (1944 ed.). Scale of both maps: 1/62500.

[3] That Hutchins was accustomed to determining latitude by "shooting the sun," is indicated by calculations in Hutchins Papers, Vol. II.

[4] On the circumferentor see section headed "Trees, Chain, Plat and Compass," chap. iii, above.

SURVEYING DIAGRAMS

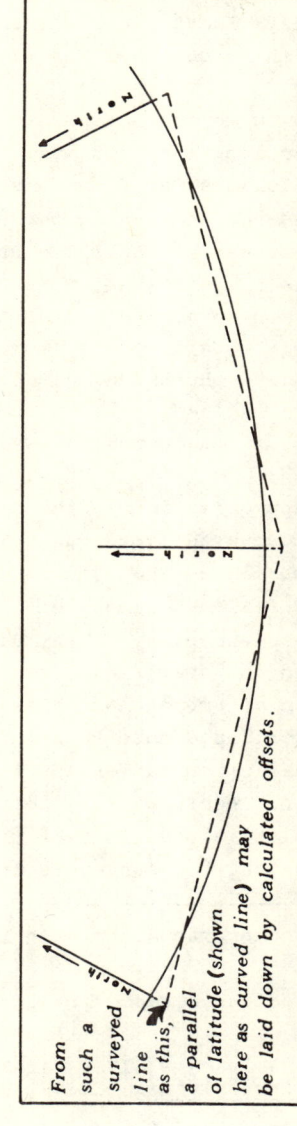

A. CONSTRUCTION OF CURVED PARALLEL OF LATITUDE UPON STRAIGHT SURVEYED LINES

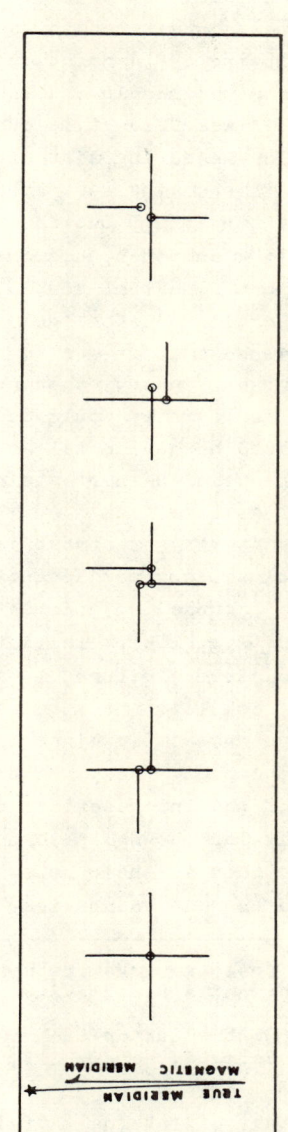

B. SELECTED TOWNSHIP CORNERS IN THE SEVEN RANGES

Fig. 13

manner. In establishing an initial line of direction, they used the circumferentor's magnetic needle. In extending a line, once begun, they seem not to have employed backsights, but to have taken a new compass reading at each advance of the instrument. And in setting off a right angle, they almost certainly read directly from the needle instead of turning the angle on the instrument. This was free-style surveying.[1]

For measuring distances, the surveyors used Gunter's chains.[2] The chains were checked for length by Hutchins, at the outset of surveying, but their results were far from consistent. It is well known that such chains altered in length, through use, but the most important source of error was probably the roughness of the terrain. "Before going a mile," wrote a later surveyor, "I discovered that it was impossible to do accurate chaining in such a broken country, where the hills were so steep it was often with difficulty they could be climbed."[3] Hutchins' directive that "the measure be horizontal as near as may be"--that is, that chaining be executed in short stair-steps up and down hill--was evidently of small effect.[4]

What with an almost casual approximation in the measurement of direction and distance, the surveyed lines generally failed, of course, to join satisfactorily at the corners of the townships (Fig. 13B). The surveyors failed to meet this problem of poor closure, in turn, in any agreed-upon way. They did not regularly complete their townships in one specified corner. They did not retrace their lines in search of error when a faulty closure occurred. They not only established more than one marker at many of the intersections of township boundaries, but they frequently left township corners open. Worst of all, they often failed to show in their notes the relative positions of these separated ends of boundaries.[5] In brief, the surveyors dramatized

---

[1] These remarks are based upon an extended examination of the survey notes for the Seven Ranges.

[2] On the Gunter's chain see section headed "Trees, Chain, Plat and Compass," chap. iii, above.

[3] Report of C. H. Van Orden, February 9, 1784, in Report of Secretary of Internal Affairs of Pennsylvania (1887), p. 403.

[4] Hutchins' only discovered reference to this manner of chaining, which had for a long time been standard practice, may be found in Hutchins to Knox, December 17, 1787 (copy), Hutchins Papers, Vol. III.

[5] These statements are principally based upon examination

the need for a standard operating procedure.[1]

Considering these technical lapses, one is tempted to look upon the survey of townships in the Seven Ranges as a source of no more than negative lessons for future public land surveying. The fact is, however, that the first surveyors made certain positive contributions. The lasting precedents which they set in perpetuating lines and corners in the field, and in recording surveys, will be given recognition in the next chapter.

The Cost of Surveying

Judging from surviving records, one concludes that all parties associated with the creation of the Seven Ranges were far more concerned with the problem of paying for the surveys than they were with the accuracy of their execution. The problem of payment is discussed below, from the point of view of the Geographer, the surveyors, and Congress, respectively.

The Geographer's pay. Hutchins, who was probably expected to produce complete returns for the first seven ranges of townships within one year of the passage of the Land Ordinance, was at first granted a generous credit on his account with the Board of Treasury.[2] But when, after nearly two years, he had submitted returns for only four ranges of townships, Congress reduced his salary, and directed that he be paid only "for such time as he may

---

of the surveyors' notes, Records of the General Land Office, National Archives, and upon a map of survey lines compiled by the author from information in the notes, from letters and diaries, and from the following topographic sheets for Ohio of U.S. Geological Survey, Topographic Map of the United States: Antrim (1911 ed.), Cadiz (1901 ed.), Cameron (1942 ed.), Carrollton (1912 ed.), Clarington (1940 ed.), Dover (1912 ed.), Flushing (1905 ed.), Macksburg (1905 ed.), Marietta (1927 ed.), New Martinsville (1926 ed.), New Matamoras (1926 ed.), St. Clairsville (1905 ed.), St. Marys (1927 ed.), Salineville (1905 ed.), Scio (1904 ed.), Steubenville (1942 ed.), Summerfield (1911 ed.), Uhrichsville (1937 ed.), Wellsville (1944 ed.), Wheeling (1942 ed.), and Woodsfield (1905 ed.). Scale: 1:62,500.

[1]Fortunately, the glaring errors committed in the Seven Ranges came to light soon enough to serve as a warning to the first Surveyor General appointed after passage of the Land Act of 1796. See Rufus Putnam to Zaccheus Biggs, Marietta, April 22, 1801, in Carter, Territorial Papers, III, 130-132.

[2]Hutchins' account was credited for $6,700.00 in June, 1785, according to Journal "C," p. 1230, Records of the General Accounting Office (Record Group 217), Fiscal Section, Legislative, Judicial and Diplomatic Records Branch, the National Archives.

be actually employed in the public service."[1] By this time, Hutchins had exhausted the public funds at his disposal, and was living on money loaned to him by friends.[2] At the end of somewhat more than three years, by which time returns on all of the Seven Ranges had been deposited with the Board of Treasury, Hutchins succeeded in collecting arrears on his salary only by directing a special appeal to Congress, in which he justified his claim against the federal government.[3]

Hutchins' account with the Board of Treasury was complicated both by the fact that he paid official as well as personal expenses out of his salary, and that he acted as bursar for the Geographer's Department. As an absorber of official expenses, Hutchins had difficulty in persuading the Board of Treasury to establish a reasonable limit to his accountability. The Board recognized certain "contingencies," for which Hutchins was reimbursed, only after trying delays.[4] As bursar, Hutchins took it upon himself to pay cash to the surveyors in advance of their earnings. To his dismay, he found the Board holding him personally responsible for these advances, and at one point he considered suing the several surveyors for nearly three thousand dollars, which the Board had refused to pass to his credit.[5] Only after a delay which imperiled relations between the Geographer and the surveyors did the Board yield on this point.[6]

Hutchins' compensation for his part in the survey of the

---

[1] Jrnls. Cont. Cong., XXXII, 129, and Papers Cont. Cong., XIX, Pt. III, 243.

[2] Hutchins to Ebenezer Sproat, New York, April 1, 1787, Hutchins Papers, Vol. II.

[3] Hutchins to Special Committee of Congress, New York, August 5, 1788 (copy), George Morgan Papers, Illinois Historical Survey, Urbana, Illinois. This letter lead to authorization of payment of back salary, recorded in Jrnls. Cont. Cong., XXXIV, 406.

[4] The extra expenditures of which Congress eventually relieved Hutchins consisted mainly of the wages of messengers. (Papers Cont. Cong., IX, 313.)

[5] Hutchins to President of Congress, New York, May 9, 1787, Papers Cont. Cong., LX, 305.

[6] The Board of Treasury recommended that Congress act to repay Hutchins for advances to the surveyors in excess of their earnings, October 2, 1787. (Papers Cont. Cong., CXXXVIII, Pt. I, 621-624.) For favorable Congressional action, see Jrnls. Cont. Cong., pp. 598-599.

Seven Ranges, after the exclusion of contingent expenses and the disentanglement of his accountability from that of the surveyors, was about five thousand five hundred dollars.[1] Of this reward he said, "It is not adequate to the unavoidable expenses in the prosecution of my Duty."[2]

<u>The surveyors' pay</u>. The surveyors were plagued by unexpectedly high overhead costs, and were restricted by a rate of pay--two dollars per mile--which barely covered their daily operating expenses. Their overhead costs were threefold in origin. First was the cost of transportation between the respective homes of the surveyors and Pittsburgh. At least three of the surveyors economized by wintering on the Ohio River, between survey seasons, but the remainder travelled to and fro annually, apparently hoping that Congress would not require them to pay for transportation from their earnings.[3] Second was the cost of provisions, and the hire of a hunter to supplement the food supply of each survey party.[4] Third was the cost of maintaining a survey crew, recruited on the frontier, during extended periods of inactivity.[5] Since the surveyors accomplished no surveying on their own account in 1785, all of their expenses took the form of

---

[1] Hutchins was paid from May 27, 1785, the day of his appointment under the Land Ordinance of 1785, to March 31, 1787, at a rate of $6.00 per day. This made his annual income about the same as that of a commissioner of the Board of Treasury. From April 1, 1787 to June 30, 1788 he was paid at a rate of $1,500.00 per year. Other officers of government under the Confederation suffered a proportionate reduction in pay, along with Hutchins.

[2] Hutchins to Don Diego Gordoqui, Pittsburgh, November 29, 1788, Facsimiles from Spanish Archives, Archivo Historico National, Legajo 3849, Apartado 1, Letter No. 306, Manuscripts Division, Library of Congress.

[3] The best single source on the travels of the surveyors is the set of accounts submitted to Congress by the surveyors for 1785. The cost of travel in that year ranged from $25.00 to $50.00, one way. (Papers Cont. Cong., XLI, Pt. IV, 305, 307, 309, 315, 317.) Various letters and journals, heretofore cited, offer information on travel in 1786 and 1787.

[4] One of the surveyors in 1785, for example, paid a hunter a wage of $0.33 a day to provide his party with meat, after having spent over $50.00 in Pittsburgh for "flour and other necessaries." (Account of Absalom Martin, Papers Cont. Cong., XLI, Pt. IV, 309.)

[5] Martin supported a crew of four men for about one month without benefiting from their services, in 1785. <u>Ibid</u>.

overhead, for the coverage of which they petitioned Congress upon their return from the field. In 1786 and 1787, the surveyors' overhead greatly exceeded their ability to pay from earnings at the established rate, and their appeals to Congress for relief were renewed. Accepting these claims, Congress paid out more money than the entire surveying enterprise was originally expected to cost, as will be shown.[1]

Overhead aside, the operating expenses for each surveyor when active in the field totalled about three dollars per day. This amount was required for the hire and support of two chairmen, an axe-man, and a packhorseman, and the rental of two or three horses.[2] Since a surveyor seldom succeeded in proceeding more than two miles per day, the need for an increase in pay above the rate of two dollars per mile was evident. Hutchins spoke out on behalf of the surveyors, and in consequence Congress once considered increasing the rate, but no action was taken during the period here under consideration.[3]

It is clear that the surveyors collected from the federal government, however belatedly, payment for both their overhead and operating expenses. Further, under both heads, their personal expenses--for food, lodging, and incidentals--were evidently covered. But if the surveyors were compensated for the time they spent in the service of the government, this fact was concealed by their accounts.[4] Repeatedly, the various surveyors appealed for extra compensation for their time on duty, but Congress consistently ignored these requests. Considering this, one is forced to conclude either that the surveyors "padded" their expense accounts to provide compensation for themselves, or that they served with-

---

[1] Congress was first induced to pay all the expenses incurred by the surveyors in 1786. See Resolution of September 25, 1786, Jrnls. Cont. Cong., XXXI, 687. Then, October 3, 1787, Congress agreed to relieve the surveyors of the costs, chiefly comprising overhead, which the surveyors had been unable to meet from their earnings in 1786 and 1787. See ibid., XXXIII, 598-599.

[2] Hutchins to President of Congress, New York, March 16, 1786, Papers Cont. Cong., LX, 237-240.

[3] Ibid., and Report of Committee on Surveyors' Pay, Papers Cont. Cong., XIX, Pt. III, 237-238. The committee favored raising the rate to $3.00 per mile.

[4] Payments to the surveyors consisted, in the last analysis, simply of a coverage of their declared expenses, which did not include the price of the surveyors' own labor.

cut any gain whatsoever.[1]

The total bill for surveying. However justified were the demands of Hutchins and his surveyors, Congress had reason to view the ultimate cost of surveying as exorbitant. If Congress originally expected seven ranges of townships to be made ready for sale within one year, as seems to be the case, the following schedule indicates the expenses which were anticipated:[2]

Geographer's salary 365 days at $6.00 per day	$2,190.00
Surveyors' earnings 1,300 miles at $2.00 per mile	2,600.00
Total	$4,790.00

Contrary to this expectation, Congress was confronted by greatly expanded financial obligations. Hutchins' salary was prevented from exceeding an amount three times the expected figure only by reducing its rate, and deducting for a leave of absence.[3] The surveyors' rate of pay was nominally held under control, but emergency appropriations, which were required to make up the surveyors' mounting deficits, more than doubled the rate, in effect. Finally, unexpected contingent expenses arose for which additional allowance had to be made. In consequence, the total bill for surveying was as follows:[4]

---

[1] For requests for compensation beyond the coverage of expenses, all of them denied, see Surveyors to President of Congress, Banks of the Ohio, August 14, 1786, Papers Cont. Cong., XLI, Pt. LX, 415-416; Surveyors to Congress, Ohio County, Virginia, November 10, 1786, ibid., pp. 431-434; William Morris to Congress, New York, February 27, 1787, ibid., XLII, Pt. V, 391-392; Surveyors to Congress, New York, September 22, 1787, ibid., XLI, Pt. IX, 461-464.

[2] The rates listed here are those which were fixed under the Land Ordinance of 1785. The number of miles cited approximates the quantity which the first seven ranges of townships were expected to contain, in the summer of 1785.

[3] For reduction of rate, already spoken of, see Jrnls. Cont. Cong., XXXII, 128-129. The leave of absence for which deduction was made was for Hutchins' tour of duty as boundary commissioner for New York and Massachusetts, in 1787. (Papers Cont. Cong., XIX, Pt. III, 242.)

[4] These figures have been checked in detail against the following records of the Board of Treasury: Journal, August 1, 1785-June 8, 1787, and Journal June 8, 1787-July 14, 1789, Records of the Bureau of the Public Debt (Record Group 53); and Journal "C" and Journal "D," Records of the General Accounting Office (Record Group 217), Fiscal Section, Legislative, Judicial and Diplomatic Records Branch, the National Archives.

```
Geographer's salary
    674 days at $6.00 per day        $4,044.00
    367 days at $1,500.00 per year    1,509.26
                                                  $ 5,553.26
Surveyors' earnings
    1,081.84 miles at $2.00 per mile                2,163.68
Extra compensation to surveyors
    For 1785                          1,673.69
    For 1786, 1787                    4,929.13
                                                    6,602.82
Contingencies                                         556.69
                           Total                  $14,876.45
```

In brief, surveying cost three times as much (and took three times as long) as Congress probably expected. On this account alone, it is not surprising that Congress questioned the wisdom of the original program for squaring off the land.

CHAPTER VIII

THE SIGNIFICANCE OF THE FIRST SURVEYS

The surveys upon which Congress placed its hopes for the retirement of the public debt, in 1785, yielded the federal government little more than one hundred thousand dollars. This amount, realized from the sale of land in the Seven Ranges, was not forthcoming until more than two years after the passage of the Land Ordinance of 1785, and it was not increased during the lifetime of that law. It was far from sufficient for even one year's payment on the principal and interest of the public debt.[1] In consequence, the laying out of the Seven Ranges must be regarded as an enterprise that failed in the accomplishment of its primary end. The present chapter, after rendering an account of Congress' attempt at money-making, proceeds to a consideration of the broader significance of the first surveys, wherein they are given credit for (1) making basic contributions to the tradition of public land surveying, (2) promoting the settlement of the West, (3) improving the mapping of the West, and (4) providing valuable sources of historical information.

The first and only sale of land in the Seven Ranges, under the Land Ordinance of 1785, was held in New York City, September 21-October 9, 1787.[2] An impatient Congress, upon receiving returns for the first four ranges of townships from Hutchins, in the spring of 1787, acted to permit the sale of this land without regard to the three remaining ranges, and abruptly cancelled the clause in the Land Ordinance that required the surveyed land to be apportioned among the several states for local sale.[3] Notice

---

[1] The scheduled payment for 1787 on the government's indebtedness to foreign creditors was about nine hundred thousand dollars, and on its debt to domestic creditors, nearly twice that much. (Estimate of Monies Requisite for the Services of the Year 1787, Papers Cont. Cong., CXIV, Pt. II, 161.)

[2] These dates appear on the record of sales for this land auction, Papers Cont. Cong., LIX, Pt. III, 135.

[3] Congress, making these changes by a resolution of April 21, 1787, simply approved a report of the Board of Treasury, for which see Carter, **Territorial Papers**, II, 24-25.

of an auction, to be held "at the expiration of five months from date," was ordered to be printed "in one of the newspapers at least of each of the states," and the Secretary of War was instructed to select land to be reserved for the satisfaction of soldiers' bounties.[1] At the ensuing sale, where Hutchins and several of his surveyors were on hand to advise prospective buyers, twenty-seven whole and fractional townships were made available for purchase, fifteen of them subject to division into square-mile sections. Land immediately bordering the Ohio River generally found a ready market, both in sections and in fractional townships, and two townships near the Ohio were sold as whole units, but buyers could not be tempted very far inland nor induced to take up all of the land near the Ohio so long as a minimum price of one dollar per acre prevailed. With less than one-third of the land spoken for, the auction was closed, despite an earlier promise that sales would "continue from day to day until the whole are sold."[2]

The subject of revenue from this sale deserves brief analysis. The total price, for thirty separate purchases, was $176,090.[3] The average price bid per acre was $1.26. Almost exactly one-half of the total sales figure was accounted for by one large speculative transaction. To stimulate sales at the auction, Congress had resolved to accept one-third of the purchase money as a down payment, if desired, and the purchasers involved in this largest transaction availed themselves of this credit privilege. Since they later failed to complete payment, the government's revenue was reduced to a sum about $60,000 below the total purchase price.[4]

---

[1] Jrnls. Cont Cong., XXXII, 226.

[2] The best single source on this land auction is in A. M. Dyer, "First Ownership of Ohio Lands," New England Historical and Genealogical Register, LXIV (October, 1910), 367-369 and LXV (January, 1911), 51-53. Dyer reproduces much documentary material, but fails to reproduce in full an important document to be cited in the next note, below.

[3] Payment was made entirely in what Hutchins called "liquidated accounts and public secretaries of the United States." (Hutchins to _____, January 10, 1788, Hutchins Papers, Vol. III.) All figures cited on this land auction are based upon "Schedule of Sales of Lands in the Western Territory," Papers Cont. Cong., LIX, Pt. III, 135-140.

[4] Of the total purchase price for this land, $88,764, only $29,669 was paid in, consisting of a one-third down payment plus surveying charges. This speculative transaction, recorded under the names of Alexander McComb and William Edgar, takes on added

And since an extra assessment against all purchasers of $1.00 per square mile for surveying was of negligible effect in covering the cost of the surveys, a further deduction of more than $14,000 for surveying must be made in reckoning the government's income.[1] From the remaining amount--little more than $100,000, as indicated at the opening of this chapter--subtraction should probably be made for part of the cost of maintaining troops on the Ohio River. The original hopes of Congress for great financial gain were obviously misplaced.

Nine years after this auction, in 1796, more land within the Seven Ranges was exposed to sale.[2] In the meantime, the Board of Treasury neglected the unsold land in the first four ranges, together with all of the townships in the last three ranges for which Hutchins submitted plats and notes in 1788, while Congress lent its support to private enterprisers who undertook their own township surveying beyond the borders of the Seven Ranges. Discussion of this later surveying will be deferred until the next chapter, to allow for a broadened consideration of the significance of surveying in the Seven Ranges, at this point.

Influence upon Later Public Land Surveying

One may say with good reason that the survey of the Seven Ranges stands as a case apart in the history of U.S. public land surveying. Here alone was surveying undertaken as an act of subdividing one of Jefferson's western states, and surveyors without contractual obligations, each representing a state, were dispatched from a common base line, in sequence, by a presiding official. Further, here alone were surveyors unconditionally released from the necessity of surveying "by the true meridian." Despite these unique characteristics, work in the Seven Ranges did much to point the way for future public land surveying.

---

interest because of evidence that Thomas Hutchins had invested in the venture. (Quattrocchi, "Thomas Hutchins," p. 263.)

[1] Only $235 was collected from purchasers to cover surveying, whereas Congress had already largely committed itself to the payment of nearly $15,000 for surveying in the Seven Ranges. See section headed "Total Cost of Surveying," chap. vii, above.

[2] See section headed "Passage of the Land Act of 1796," chap. x, below. In July, 1788, in a supplement to the Land Ordinance of 1785, Congress authorized further sale of land in the Seven Tanges, but no sales ensued. For this law see Jrnls. Cont. Cong., XXXIV, 306-310.

First, certain negative lessons were effectively taught here. The impracticality of the scheme for engaging "gentlemen surveyors," apportioned among the several states, was quickly demonstrated. The collecting together of surveyors for directly supervised work was shown to be an unwieldy arrangement. The need for limiting the financial liability of the federal government was made evident. And experience pointed to the desirability of adopting standard field procedures, and of restoring the "true meridian" clause to the law governing surveys. In brief, the survey of the Seven Ranges was instructive by virtue of its errors.[1]

Second, certain positive contributions were made by the surveyors of the Seven Ranges, in marking and making a record of their surveyed lines. Told no more by the Land Ordinance about leaving signs in the woods than that "the lines . . . shall be plainly marked by chaps on the trees," the surveyors set a lasting precedent in their way of marking "line trees"[2] and corner points,[3] and in their practice of witnessing corner posts through the use of "bearing trees."[4] Again, told simply that "the lines

---

[1] It might be argued that Thomas Hutchins was deploring the shortcomings of field practice in the Seven Ranges specifically rather than the faults of the rectangular survey system in general, when he proposed the abandonment of rectangular surveying, in 1788. See Hutchins to Committee of Congress, New York, March 5, 1788, Papers Cont. Cong., LXXVIII, Pt. XII, 541.

[2] A "line tree," or tree standing directly in the line of sight between corners, was marked with notches cut by axe into the sides of the tree facing the line. Of these notches, or narrow bite-like marks, all surveyors but one (Ludlow) made only one to a side. Ludlow made two, and his example has been followed down to the present day. (U.S. Department of the Interior, Bureau of Land Management, Manual of Surveying Instructions, 1947, pp. 239-240.)

[3] At corner points, that is, at one-mile intervals along township boundaries and at the corners of townships, the surveyors initiated the practice of setting marked posts where there were no trees and, where a tree might stand precisely at a corner, of cutting identifying marks in a "blaze," or broad, flat scar on the tree. For modern practice based upon these precedents see ibid., pp. 247, 266-270. The posts, of course, were simply of native woods, such as white oak, walnut, dogwood and ash. The identifying procedure had no consistency.

[5] Near each corner point the surveyors selected two trees marking with the axe, making a record in their notes of which the following (from the survey notes of Ebenezer Sproat, for the Fourth Range) is an example: "From the Mile post a spanish Oak 14 Inches diameter bears N83°15'E 30 1/2 links distant with 6 notches and a blaze under them--and a white Oak 20 Inches diameter

... shall be exactly described on a plat," the surveyors passed on to their successors the convention of drawing plats at a scale of two inches to the mile,[1] the practice of submitting not only plats but survey notes, as a part of the official record, and the custom of assembling these notes, in transcribed form, so that accounts of the four boundaries of each township could be read in sequence.[2] These precedents were by their nature subject to transmission without personal contact between the responsible surveyors and their successors.

Finally, a chain of personally communicated surveying experience began in the Seven Ranges. Three men engaged in surveying there not only carried their experience into parts of the Ohio country controlled by private land companies, but they personally bridged the temporal gap between the laying out of the Seven Ranges and the surveying of public lands which began again in 1797--a resumption to be discussed toward the end of this study.[3] In 1797, these men established contact with a tradition which has lived on, through successive generations of public land surveyors, to the present day.

To conclude, field work in the Seven Ranges, isolated though it was in many ways, exerted a lasting influence upon subsequent surveying. In justice, it must be placed alongside the Land Ordinance of 1785 as a separate part of the foundation of the American rectangular land survey system.

---

bears S 20°30'E 50 links distant, marked in the same manner." For modern practice based upon this precedent see ibid., pp. 272-280.

[1] The present location of the plats in the National Archives has been indicated. For evidence that their scale of two inches to the mile (often expressed as "1 inch equals 40 chains") persists to the present day, see ibid., p. 401. These plats, it should be added, failed to show the relationship between true and magnetic north, concealed inaccuracies in the measurement of distance and direction, and completely ignored the existence of open and double corners, of which there were many.

[2] The notes, whose place of custody in the National Archives has been indicated, locate and identify not only bearing trees, as has been shown, but also line trees. Further, they locate stream crossings, and describe in general for each mile the soil, lay of the land and timber cover. That these are not the original notes from which the plats were made is demonstrated by the fact that the plats sometimes show features not mentioned in the notes.

[3] The three men were Israel Ludlow, Absalom Martin and John Mathews.

Contribution to the Opening of the Northwest

We should remember that the Geographer and surveyors were expected, by the authors of the Land Ordinance, to serve the interests of settlement by doing more than square off the land preparatory to its sale. They were expected to send back intelligence reports from the frontier. Accordingly, the plats and notes which Hutchins submitted to the Board of Treasury in 1787 and 1788 were designed not only to locate the lines of survey, but also to give a picture of the lay of the land, to identify forest cover, and to suggest the land use capabilities of the respective townships in the Seven Ranges.[1] Plats and notes for the first four ranges were displayed for inspection at the land auction in New York. Not content with this, Hutchins and some of his surveyors appeared personally at the sale, to offer their advice to prospective purchasers.[2] In consequence, knowledge gained in the course of surveying affected the selection of land within the Seven Ranges; but, as will now be explained, the principal contribution which such knowledge made to the opening of the Northwest pertained to settlement beyond the Seventh Range.

The major beneficiaries of intelligence gathered in the course of surveying the Seven Ranges were the organizers of the Ohio Company of Associates, who contracted to buy a large tract of land immediately west of the Seven Ranges, a few days after the public auction in New York was closed.[3] The story of the way in which this group benefited from the national surveys begins in 1785, when General Benjamin Tupper, by adopting the role of sur-

---

[1] The plats and notes in no case approached the detailed informativeness of Hutchins' initial report on the Geographer's Line. Simpson and Ludlow, in particular, were very sparing in their note-taking. An army officer who had been assigned to protect the surveyors and who was acquainted with the content of the survey notes, advised a friend that investors could easily be "bit" unless they had the direct advice of the surveyors themselves. (Captain Jonathan Heart to Major William Judd, Camp near Grave Creek, November 15, 1786, Jonathan Heart Papers, Western Reserve Historical Society.)

[2] Hutchins to Committee of Congress, New York, August 5, 1788, George Morgan Papers, Illinois Historical Survey; and Hutchins to _____, January 10, 1788, Hutchins Papers, Vol. III. In the latter letter Hutchins made plain his willingness to advise speculators.

[3] For text of contract, signed October 27, 1787, see Hulbert, *Records of the Ohio Company*, I, 29-37.

veyor for Massachusetts, found an opportunity for learning at first hand about the route to Pittsburgh and the country downstream from that settlement for a distance of about forty miles. The Ohio Company had not yet been formed, but its prospective organizers, represented by Tupper, were known to be contemplating the founding of a colony. By the summer of 1786, provisional articles of the Ohio Company had been drawn up, at a meeting in Massachusetts, and federal surveying began to look as though it were specifically meant to serve the exploratory interests of this association.[1] Five Ohio Company men, including Tupper, appeared among the federal surveyors this year,[2] and one of them, Winthrop Sargent,[3] detached himself from the rest to reconnoitre the district on the lower Muskingum River which the Company was soon to apply for, in Congress.[4] The fact that this tract of land lay west of the Seven Ranges should not lead one to suppose that Sargent thought of it as an area beyond the scope of federal surveying. Rather, he viewed it at this time as land included within the breadth of thirteen ranges of townships scheduled for survey in 1786.[5] By 1787, however, the outlook had changed. With only

---

[1] For preliminary articles of association, approved March 3, 1786, see *ibid.*, pp. 4-11.

[2] These men were, in addition to Tupper, Ebenezer Sproat and Winthrop Sargent, who came west to serve as surveyors; and Anselm Tupper, son of Benjamin Tupper, and John Mathews, nephew of Rufus Putnam, both of whom came as chainmen. Mathews, whose Journal has been of great value in the preparation of the present study, was introduced to Hutchins by Putnam as "a young gentleman who has made considerable improvement in the art of surveying, [and who] comes out a volunteer and means to tarry in the country." (Putnam to Hutchins, Rutland, Massachusetts, July 7, 1786, Hutchins Papers, Vol. III.)

[3] Sargent, quite aside from the historical value of his Journal, frequently cited in the present study, was the most important public figure involved in the survey of the Seven Ranges. Soon to take a leading role in the organization of the Ohio Company, Sargent later became Secretary of the Northwest Territory, in which office he served as de facto governor for more than half the time, 1788-1798. From 1798 to 1801 Sargent served as Governor of the Mississippi Territory.

[4] Diary of Winthrop Sargent, entries for July 23 to August 1, 1786. The significance of this visit was perhaps first emphasized in Benjamin H. Pershing, "Winthrop Sargent: A Builder in the Old Northwest" (Unpublished Ph.D. dissertation, Department of History, University of Chicago, 1927), pp. 12-13.

[5] Sargent to Samuel Parsons, Fort Harmar, August 1, 1786, Samuel Parsons Papers, Western Reserve Historical Society.

seven ranges of townships begun by the national surveyors, the Ohio Company, impelled by an apparently new determination to obtain land in a single block, threw its influence behind a move to halt any further extension of national surveying to the west.[1] Deciding to apply to Congress for a direct grant of land, directors of the Company declared, "We . . . wish, if possible, to have our eastern bounds on the seventh range of townships."[2] The Company succeeded in obtaining a grant with this eastern boundary (Fig. 14), and went on to conduct its own township surveying, as the following chapter of this study will disclose.[3]

In a well known advertisement of its new purchase, the Ohio Company drew freely upon the opinions and observations of its representatives who had engaged in the survey of the Seven Ranges, a procedure justified by the fact that the Ohio Company lands comprised a continuation of the Allegheny Plateau country wherein the Seven Ranges lay.[4] By way of further reliance upon the federal surveys, this same advertisement exploited the reputation of Thomas Hutchins, by including his testimonial that descriptions appearing therein were "judicious, just and true," and consistent with "observations made by me."[5]

---

[1] For resolution passed at the second general meeting of the Ohio Company, March 2, 1787, calling for immediate application to Congress for a large private purchase of land, see Hulbert, Records of the Ohio Company, I, 12.

[2] Quotation from Rufus Putnam and Manasseh Cutler, writing to Winthrop Sargent late in May, 1787, ibid., p. liii.

[3] Although no contract with the Ohio Company was signed until October 27, 1787, the Company's application had been referred by Congress to the Board of Treasury "to take order," July 27, 1787. (Jrnls. Cont. Cong., XXXIII, 429.) The Board simply delayed action until the auction of Seven Ranges land in New York had closed.

[4] The advertisement, titled "An Explanation of the Map Which Delineates That Part of the Federal Land Comprehended between Pennsylvania West Line, the Rivers Ohio and Sioto and Lake Erie . . . ," may be found in Philip Lee Phillips, The First Map and Description of Ohio, 1787, by Manesseh Cutler: A Biographical Account (Washington: W. H. Lowdermilk and Co., 1918), pp. 25-41. For descriptive passages patently based upon observations made in the area of the Seven Ranges, see ibid., pp. 29-30.

[5] Ibid., p. 26. An additional example of publicity derived from the surveys was Winthrop Sargent's "List of Forest and Other Trees Northwest of the River Ohio," Memoirs of the American Academy of Arts and Sciences, II (1793), 156-159.

Nor did the services rendered to the Ohio Company by the federal surveys end here. In the course of surveying in the Seven Ranges, the Army's influence had been brought down the Ohio to the mouth of the Muskingum River, where the Ohio Company's first settlement would soon be made; the Indians had been introduced to the kind of surveying which the Ohio Company would be continuing; and the squatter population of the Ohio country had been confronted by the determination of Congress to deny the right of preëmpting land by "tomahawk claim," a legal position which the Ohio Company was resolved to perpetuate.[1]

If the founding of Marietta at the mouth of the Muskingum River (Fig. 14) by the Ohio Company, in April, 1788, is to be accepted as the beginning of organized American settlement in the Northwest,[2] then the Seven Ranges should be recognized with appropriate honor as the bridgehead which made the success of this pioneer venture possible.

A Service to Mapping

From the time that the idea of dividing the public domain into squares was first officially proposed, in the Jefferson-Williamson plan of 1784, it must have been obvious that resultant surveying would prove to be a great boon to the mapping of the West. A simple and effective framework was promised, which would permit the proper placement of landscape features on a map of the public lands surveyed. The first area to be favored with this framework, of course, was that of the Seven Ranges.

For the sake of mapping, a more fortunate place for beginning the surveys than that selected by Congress could hardly have been imagined. By beginning at a point on Pennsylvania's western boundary, the surveys were placed in contact with the eastern seaboard through a line of traverse whose accuracy was unexcelled at this time.[3] As we have seen, the standards of accuracy observed

---

[1] With respect to preëmption, it should be understood that the Ohio Company was prepared to convey land-titles only to its own shareholders. In the very month that the Ohio Company made its first settlement, Congress renewed its denial of the right of roving pioneers to take up land at will on the public domain. (Resolution of April 24, 1787, in Jrnls. Cont. Cong., XXXII, 231.)

[2] The arrival of the party of founders at Marietta is recorded in Journal of John Mathews, Hildreth, Pioneer History, pp. 191-192; entries for April 7 and 8, 1788. The party's members are listed in Hulbert, Records of the Ohio Company, I, 24, n.

[3] This connecting line, simply consisting of parts of

in the survey of the Seven Ranges left much to be desired, but no amount of carelessness in the laying out of townships could destroy this initial advantage. Furthermore, despite the approximations which Hutchins and his surveyors allowed themselves, the township grid which they produced was obviously superior to other means of control upon which mapping of the Northwest had hitherto been based.[1]

The primary features of the Ohio country which called for improved representation were the watercourses. In keeping with the requirements of the Land Ordinance, township plats which Hutchins submitted to the Board of Treasury showed streams generally at the points of their intersection with township boundaries, and, going beyond the Land Ordinance, showed the Ohio River throughout its course along one edge of the Seven Ranges.[2] Even before all seven ranges were reported upon, this drainage content of the plats was exploited in a privately published map.

The first man to incorporate stream lines from the public land surveys into a map was Manasseh Cutler, a promoter of the Ohio Company who issued the advertisement on behalf of that company which, as has been said, drew upon the verbal descriptions of public land surveyors. In 1787, Cutler issued a map to accom-

---

Pennsylvania's border, had been run from the Delaware River westward for a distance of nearly two hundred fifty miles by Mason and Dixon, 1763-1767, and had been continued westward in 1784 by Rittenhouse, Hutchins, Ellicott and other leading astronomical surveyors of the time. From the southwestern corner of Pennsylvania northward to the initial point of the public land surveys, the connecting line was laid down in 1785, as described under "Establishment of the Beginning Point," chap. vi, above.

[1] Based ultimately upon latitude readings and compass courses accumulated from many expeditions in the Northwest, two maps other than Hutchins' map of 1778 were generally relied upon at the time that public land surveying began. Both published in 1755, they were Lewis Evans' A General Map of the Middle British Colonies, in America, and John Mitchell's A Map of the British and French Dominions in North America. See Susan M. Reed, "British Cartography of the Mississippi Valley in the Eighteenth Century," Mississippi Valley Historical Review, II (September, 1915), 213-224.

[2] Those plats for the Seven Ranges which showed segments of the Ohio River served to correct the earlier work of Hutchins himself. In 1766, as a British engineering officer, Hutchins had charted the entire course of the Ohio River from a boat. See Beverley W. Bond, Jr. (ed.), The Courses of the Ohio River Taken by Lt. T. Hutchins Anno 1766 and Two Accompanying Maps (Cincinnati: Historical and Philosophical Society of Ohio, 1942). The results of this charting appeared in Hutchins' map of the West, of 1778.

pany this advertisement, showing the eastern half of the present state of Ohio. Imagination and earlier maps obviously supplied the greater part of Cutler's data, but one small area of his map was strikingly accurate. This was the area covered by the first four ranges of federal townships, whose survey had been reported upon by the time the map was compiled. Here, the Ohio River and its right-bank tributaries appeared in a properly rectified form, but immediately beyond this tract the Ohio began to deviate measurably from its true course, and other drainage lines also suffered misplacement. The "latest information," upon which Cutler declared that his work was based, permitted no further extension of accurate mapping.[2]

Further exploitation of the results of public land surveying for mapping purposes waited until 1796, when Mathew Carey published a map titled, "The Seven Ranges of Townships," in connection with the land sales which were resumed in that year.[3] This careful little representation surpassed Cutler's work not only in drawing upon data for all of the Seven Ranges, but in showing in great detail the drainage pattern developed by the federal surveyors. It contributed directly to a successor map of broader scope, which will be given recognition in a later chapter devoted to an appreciation of surveying as of 1800.

The Production of Historical Evidence

Today, more than a century and a half after their drafting, the original notes and plats for the Seven Ranges may be examined at the National Archives, Washington, D.C.[4] It would be

---

[1] This map, titled "A Map of the Federal Territory from the Western Boundary of Pennsylvania to the Scioto River Laid Down from the Latest Informations . . . ," is reproduced in Phillips, The First Map of Ohio, foll. p. 41.

[2] Hutchins had submitted returns to the Board of Treasury on the first four ranges of townships, including a general plat of the area (now lost), in April, 1787. (Hutchins to President of Congress, April 18, 1787, Papers Cont. Cong., LX, 301.)

[3] Mathew Carey, Plat of the Seven Ranges of Townships Being Part of the Territory of the United States N.W. of the River Ohio. The map is reproduced in Elroy M. Avery, A History of the United States and Its People (7 vols.; Cleveland: The Burrows Brothers, 1904-1910), VI, foll. 406.

[4] The original notes and plats, as previously noted, are located in Records of the General Land Office (Record Group 49), Cartographic Records Branch, the National Archives. They should

misleading to suggest to prospective researchers that these records comprise a rich historical record, yet an appreciation of the general significance of surveying in the Seven Ranges would not be complete without recognition of the fact that the notes and plats have historical value. In the following paragraphs, consideration is given to five subjects of known interest to historical researchers, with respect to which Hutchins and his surveyors produced documentary evidence.

Forests. Marked trees, as the reader has been informed, were left behind in the field by the surveyors, to commemorate the location of township boundaries. By recording in their notes the species, diameter and position of each of these trees, and by occasionally identifying collectively trees which came within their view, the surveyors made their most effective contribution to historical knowledge. On the basis of these notes, a botanist in recent years has been able to reconstruct, in a general way, the original forests of the Seven Ranges.[1]

Trails. Knowing that several trails antedated surveying in the area of the Seven Ranges, the present author hoped that their courses, like those of streams, would be indicated in the notes and plats by a record of their points of intersection with township boundaries. Curiously, although the surveyors themselves occasionally used these trails in making their way through the wilderness, they noted such intersections in only three places.[2] It remained for surveyors who further subdivided the Seven Ranges about twenty years later to develop a picture of the trails which criss-crossed the area at that time.[3]

---

be distinguished from the following records deposited with them, which are derivative and of a later date: (1) a calf-bound volume measuring 16 1/2 inches by 10 1/2 inches, containing an incomplete copy of field notes, titled on backstrip "Field Notes of the Surveys of the Seven Ranges of Townships Surveyed in Conformity to an Ordinance of Congress of May 20th 1785"; and (2) a set of loose sheets, each measuring 15 1/2 inches by 25 inches, and each exhibiting a township plat together with abstracts from the survey notes.

[1] For this reconstruction, which covers not only the Seven Ranges but all of the State of Ohio, see Paul B. Sears, "The Natural Vegetation of Ohio, I: A Map of the Virgin Forest," Ohio Journal of Science, XXV (May, 1925), 139-149.

[2] Plats show an intersection of trail and boundary along the east side of Township Nine, and the west side of Township Eight and Ten, Fifth Range.

[3] See plats for townships of the Seven Ranges, "Ohio Vol. 3," also titled "Record Book No. III for the Secretary of the

Sites of human habitation. To judge by the plats and notes alone, no sites of human habitation were encountered in the course of surveying the Seven Ranges. But from other sources one learns that at least five squatters' cabins, each with a surrounding clearing in the forest, were found on or near the lines of survey. Mention of this fact provides an opportunity for pointing out that the records of historical value which Hutchins and his surveyors produced were not confined to the official notes and plats. In this instance, a letter and three diaries serve to afford us information which the official records lack.[1] Similar supplementary material, it is suggested, should be consulted by researchers interested in other parts of the public domain.

Place names. Place names are represented in the survey notes for the Seven Ranges by a single type, that of names for watercourses. It should be understood that the surveyors--like their successors elsewhere in the public domain, to the best knowledge of the present author--did not confer names, but simply applied the nomenclature already in use among local frontiersmen.[2] The survey notes offer to the student of toponymy, therefore, documentary evidence of place names in use at a definite date, prior to general settlement.[3] Within the Seven Ranges, the notes

---

Treasury U.S.," Records of the General Land Office, Cartographic Records Branch, the National Archives. Archer B. Hulbert, when professor of American History at Marietta College, looked forward to using survey records for the delineation of trails in the area of the Seven Ranges. (Hulbert, Ohio in Time of the Confederation, p. 166, n.) This he did not do. He had already written on the Indian trails of Ohio, but without benefit of survey records. See his "The Indian Thoroughfares of Ohio," Ohio Archaeological and Historical Publications, VIII (1900), 263-295.

[1] "A Brief Account of the Soil and Timber . . . ," in Hutchins to President of Congress, New York, December 27, 1785, Papers Cont. Cong., LX, 229-236; entries for August 8 and 15, 1786, Journal of Thomas Hutchins, Hutchins Papers, Vol. II; entry for September 24, 1786, Diary of Winthrop Sargent, Sargent Papers; entry for September 4, 1787, Journal of John Mathews, in Hildreth, Pioneer History, p. 187.

[2] This simple recording function stands in contrast to the practice of surveyors elsewhere, for example, in Connecticut's Western Reserve, where surveyors were responsible for the bestowal of many place names.

[3] An amusing example of fidelity in the reporting of local names appears in the survey notes of Absalom Martin of New Jersey. Martin speaks of crossing "a brook called Island Creek," in his notes for Township Six, Second Range. The incidence of the term creek as a southernism in conflict with the term brook is noted in H. F. Raup, "The Names of Ohio Streams," Ohio Conservation Bulletin, XX (July, 1956), 10-11.

provide such evidence for seven, and letters and diaries identify three more, watercourses.[1]

<u>Land evaluation</u>. One is tempted to dismiss the surveyors' remarks, in their notes, on soil fertility, the quality of mineral resources, and the suitability of land to various crops, because of their superficiality. Taken together, however, these observations sketch a picture of the area concerned, as evaluated by contemporary standards. No assessment of the Seven Ranges, based upon the value judgments of the surveyors, has been attempted by a twentieth century researcher, so far as is known, but equivalent appraisals of other parts of the public domain are being compiled at the time of writing.[2]

---

[1] These are the names: Captina, Cross, Island, Little Yellow, McMahons, Sandy, Short, Wheeling, Wills and Yellow Creeks.

[2] These appraisals are being prepared in connection with suits filed under the Indian Claims Act of 1946. Both in support of and in opposition to claims by Indian tribes that they were not justly compensated by the United States when lands were taken from them under various treaties of cession, the survey records are being used to help establish the value to the white man of these lands at the time of their acquisition.

CHAPTER IX

PRIVATE SURVEYING UNDER THE LAND ORDINANCE OF 1785

Beginning in the spring of 1787, Congress cautiously developed a new land policy for the Northwest, without entirely abandoning the Land Ordinance of 1785. Persuaded that surveying then in progress under the direction of Thomas Hutchins was hopelessly slow, Congress first acted to release for sale the four ranges of townships already surveyed,[1] and then admitted to consideration certain proposed alternatives for the disposition of public lands. On the one hand, the Secretary of War urged upon Congress a plan for reserving some large tract solely for the satisfaction of soldier bounties,[2] and on the other hand, groups of investors bid for approval of their schemes for acquiring and settling selected parts of the public domain.[3] With an outlook dominated by the idea of securing revenue, Congress paid first heed to the latter petitioners, and, in anticipation of the emigration which these organizers would promote, passed a new Ordinance for the government of the Northwest, calculated to secure more effective control over the inhabitants than that provided for in Jefferson's Ordinance of 1784.[4] While this Ordinance was under debate, negotiations with the Ohio Company of Associates were drawing to a successful conclusion, and on July 27 Congress directed the Board of Treasury to draw up a contract with that group and certain associated persons for the conveyance of a tract

---

[1] Resolution of April 21, 1787, cited at opening of chap. ix, above.

[2] Secretary of War to President of Congress, April 26, 1787, in Carter, Territorial Papers, II, 27-28.

[3] The leading petitioners at this time were Nathaniel Sackett and associates, who desired the territory between the Muskingum and Scioto Rivers (Hulbert, Ohio in Time of Confederation, pp. 114-124), and Samuel Parsons, who sought an adjacent tract on behalf of the Ohio Company (Hulbert, Records of the Ohio Company, I, li-lii).

[4] This was the Northwest Ordinance, passed July 13, 1787. For earlier mention of this law, see above, pp. 16, 33, 35.

of land immediately west of the Seven Ranges (Fig. 14).[1] Shortly thereafter, John Cleves Symmes of New Jersey applied for a grant of land on terms similar to those secured by the Ohio Company, and on October 2 Congress directed the Board of Treasury to meet Symmes' request.[2] This second grant was located on the Ohio River, in the southwest corner of the present state of Ohio. Lying between the Great Miami and Little Miami rivers (Fig. 14), it came to be known as the Miami Purchase.[3]

By way of completing the new program for the Northwest, Congress voted to reserve two large bodies of land, one immediately north of the Ohio Company's grant and the other at the mouth of the Ohio River, "for the purpose of satisfying the military bounties of the late Army."[4] Congress also took action to assure continued military protection along the Ohio, and authorized the new Governor of the Northwest Territory to treat with the Indians for cessions which, it was hoped, would open up all lands from the Pennsylvania border to the Mississippi River, between the Ohio River on the south and, on the north, the latitude of the Geographer's Line.[5] Finally, Congress announced the availability of ungranted lands in this broad sweep of territory to any purchasers willing to take up "not less than one Million of Acres in One body."[6] An era of large-scale private colonization seemed to be at hand.

As is well known, the colonization schemes of only the Ohio Company and the Symmes group came to fruition. The lands purchased by these two associations required surveying, and with this

---

[1] Jrnls. Cont. Cong., XXXIII, 429.

[2] Ibid., p. 594. Symmes' petition appears in Carter, Territorial Papers, II, 70-71.

[3] The final grant, as shown, Fig. 14, is only a fraction of the million acres which Symmes originally hoped to obtain.

[4] Resolution of October 22, 1787, in Jrnls. Cont. Cong., XXXIII, 695-696.

[5] For official authorization of Governor Arthur St. Clair, see ibid., pp. 611, 696. Privately communicated instructions from Congress to St. Clair may be found in Carter, Territorial Papers, II, 78-79.

[6] Resolution of October 22, 1787, in Jrnls. Cont. Cong., XXXIII, 701. This resolution, which provided for grants in states of the Northwest other than that enclosing the Ohio Company's purchase, was precipitated by a request for a large grant by Royal Flint and associated speculators (Carter, Territorial Papers, II, 75-76).

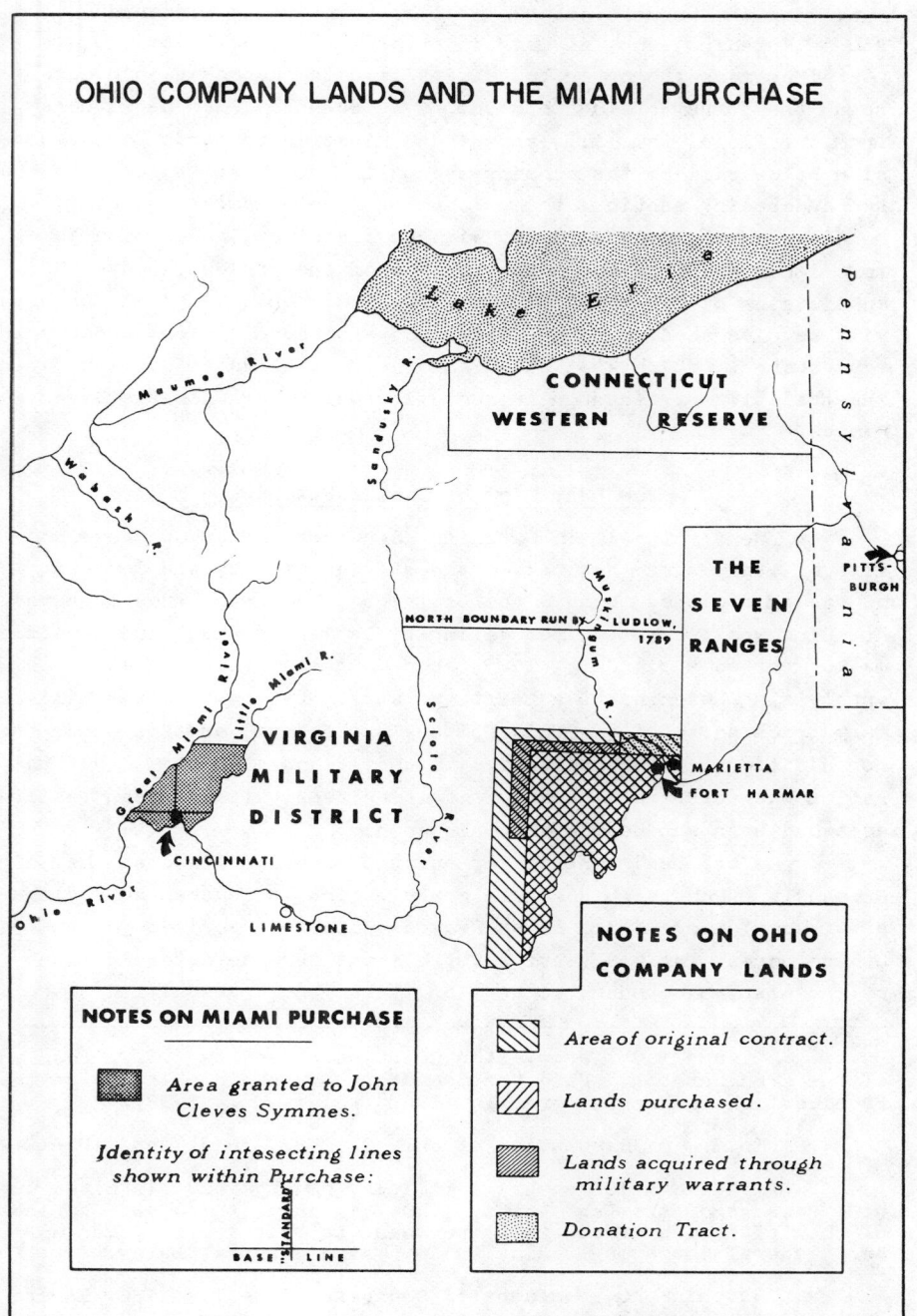

Fig. 14

surveying the present chapter will be principally concerned. Thomas Hutchins, who, as Geographer of the United States, was expected to mark the outer boundaries of both purchases, had hardly begun the survey of the Ohio Company's boundary when he withdrew from the field, fell ill, and died. First in the order of discussion below will be the closing events in Hutchins' career. In the two succeeding sections the posthumous completion of Hutchins' boundary-surveying assignment will be described, and--which is more important from the point of view of the present study--the subdivision of the Ohio Company's territory and the Miami Purchase will be discussed. It was this process of subdivision, under private control, which kept alive and, as we shall see, developed, the tradition of rectangular surveying which began in the Seven Ranges.

The Last Days of Thomas Hutchins

The first call on Hutchins' services, after Congress had altered its program for the Northwest, was made by the Secretary of War, who wished to have the newly reserved Army lands surveyed.[1] Hutchins was prevented from acting by lack of funds,[2] and here the matter rested until July, 1788, when Congress passed a resolution authorizing Hutchins to appoint two surveyors, one for the military tract north of the Ohio Company's purchase, and the other for for additional military reserve in the Illinois country,[3] but no action was taken until nearly a decade later, after Indian resistance had been subdued in the Ohio country.[4]

Before beginning to lay out the exterior lines of the Ohio Company's purchase--his single contribution to federal surveying after completion of the Seven Ranges--Hutchins received yet another request for his services. This was for the determination of the western boundary of New York.[5] The request rattled a skele-

---

[1] "I applied early after Congress passed the . . . resolve of the 22nd October," said the Secretary of War in a letter to the President of Congress, March 10, 1788 (ibid., pp. 95-98).

[2] Ibid.   [3] Resolution of July 9, 1788, ibid., pp. 122-124.

[4] A law of 1796 covering military bounty lands is cited under "Passage of the Land Act of 1796," in chap. x, below. Notice of the initiation of surveying pursuant to this law is given under "Federal Surveying Begins Again," also chap. x, below.

[5] Hutchins to President of Congress, June 16, 1788, Papers Cont. Cong., IX, 335.

ton in Hutchins' closet, so to speak, since its fulfillment would have required him to deal directly with the British government in Canada.[1] Hutchins, it was believed, was still regarded with disfavor, since he had deserted the British service in time of war.[2] Whether for this reason or not, Hutchins delegated the task of surveying the boundary to Andrew Ellicott, last mentioned in this study as a commissioner who assisted in running the western boundary of Pennsylvania.[3]

In midsummer of 1788, Hutchins was free to turn his attention to the Ohio Company's purchase, though he tarried in New York until he had arrived at a satisfactory settlement of his account with the Board of Treasury for superintending the survey of the Seven Ranges.[4] Then, late in August, he set out for the West.[5] Reaching Pittsburgh in mid-September, Hutchins soon set off down the Ohio River, and, after a few days' travel alongside the Seven Ranges, arrived at Fort Harmar.[6] Across the Muskingum River from Fort Harmar stood Marietta (Fig. 14), where settlement had been initiated by the Ohio Company six months earlier.

At Fort Harmar, Hutchins' job was to organize a survey of

---

[1] As indicated, p.   , above, the meridian bounding New York on the west was to be aligned with the western tip of Lake Ontario. The accomplishment of such an alignment called for a traverse across Canadian soil (Fig. 3), and Hutchins accordingly drafted a letter to Lord Dorchester, Governor of Quebec, for the approval of Congress (Papers Cont. Cong., LX, 331).

[2] Quattrocchi, op. cit., pp. 198, 267.

[3] Ellicott later reported that he received instructions from Hutchins to begin the New York boundary survey, July 1, 1788 (Ellicott to John Page, New York, June 1, 1789, Applications for Office under President Washington).

[4] A contentious correspondence between Hutchins and the Board of Treasury, July 19-August 2, 1788, may be found in Papers Cont. Cong., LXXVIII, Pt. XII, 549-559. Only after Congress specifically ordered payment did the Board yield in its resistance to Hutchins' claims.

[5] Hutchins to President of Congress, New York, August 15, 1788, Papers Cont. Cong., LX, 339. Congress, by this time, had extended Hutchins' term of office to May, 1790 (Jrnls. Cont. Cong., XXXIV, 180), had given him freedom of choice with respect to surveyors (ibid., p. 331), and had authorized a credit of $4,000 to his account with the Board of Treasury (ibid., p. 389).

[6] Hutchins to President of Congress, Fort Harmar, October 3, 1788, Papers Cont. Cong., LXXVIII, Pt. II, 559.

the following boundaries: (1) the course of the Ohio River from the western limit of the Seventh Range down to the mouth of the Scioto River, (2) the course of the Scioto from its mouth to a point about eighty miles upstream, and (3) a line directed due east from this point on the Scioto to the western boundary of the Seventh Range (Fig. 14).[1] While in New York, Hutchins had posted letters westward to Israel Ludlow and Absalom Martin, two veterans of surveying in the Seven Ranges, requesting their assistance.[2] Rendezvousing with Ludlow and Martin at Fort Harmar, Hutchins proceeded with an Army escort of nearly fifty men to the mouth of the Scioto, there to begin surveying. At the outset, Hutchins took the latitude of the mouth of the Scioto, and then, after surveying only twelve miles of the Ohio River, left the continuation of the work to his two deputies. The weather had impeded progress in surveying, and the prospect for early completion of the boundary assignment was clouded by a shortage of troops, and by threatening reports of Indian activity to the north.[3] Returning to Pittsburgh, Hutchins wrote an optimistic report to Congress,[4] but his words concealed the feelings of an ailing and excessively frustrated man.

To judge by the tone of his report to Congress, one would suppose that Hutchins was ready to turn next to the survey of boundaries of the Symmes purchase, as a matter of course. He had no such intention. Writing secretly from Pittsburgh to Diego de Gardoqui, Spanish emissary to the United States, he had already proposed abandoning his position as Geographer of the United States in exchange for a comparable office under the Spanish Governor at New Orleans.[5] An understanding of this drastic move on

---

[1] These were the outlines of a tract set aside by Congress by a report in Congress of July 23, 1787 (Jrnls. Cont. Cong., XXXIII, 399-401). The northern boundary was to connect with the northwest corner of Township Ten, Seventh Range.

[2] Quattrocchi, op. cit., p. 272.

[3] Ibid., pp. 272, 274; Harmar to Hamtramck, Fort Harmar, October 13, 1788, in Smith, St. Clair Papers, II, 92; and Hutchins to President of Congress, Pittsburgh, January 25, 1789, Papers Cont. Cong., LXXVIII, Pt. II, 563-564.

[4] Ibid.

[5] Hutchins to Don Diego Gardoqui, Pittsburgh, November 29, 1788, Facsimiles from Spanish Archives, Archivo Historico Nacional, Legajo 3849, Apartado 1, Letter No. 306, Manuscripts Division, Library of Congress.

Hutchins' part requires that the following four points be considered:

(1) Hutchins was impatient to return to that exploring and charting of rivers to which he owed his original reputation as an authority on the West. Twice he had proposed to Congress that he be commissioned to render a map of the course of the Ohio River and its major tributaries from the north, this map to be accompanied by "an Account of their Navigation and a general description of the Soil, Timber &c of the Country through which they run."[1] Frustrated in this desire, he hoped for satisfaction from the Spanish. "Should [the Governor at New Orleans] think it necessary to have Plans made of all the Rivers communicating with the Mississippi, with a Description of the Country . . . through which they flow," Hutchins wrote to Gardoqui, "it would add much to my happiness to be honored with Instructions for the accomplishment of so important a work."[2]

(2) Hutchins had been sorely tried in his dealings with the Board of Treasury. In the summer of 1788, he had collected his just reward for the survey of the Seven Ranges from the Board only with great difficulty.[3] Earlier, he had been prevented from obliging the Secretary of War in his desire for the survey of military lands by the Board's refusal to advance money.[4] More recently, his plans for the survey of certain small tracts granted by Congress to a group of Delaware Indians had been defeated by a lack of funds,[5] and the survey of the boundaries of the Ohio Company's purchase was attended by financial difficulty.[6] Hutchins hoped for more generous treatment at the hands of the Spanish government.

(3) Hutchins' activities in the Northwest had been adversely affected by real or threatened Indian resistance ever since his first attempt at surveying in the Seven Ranges, in 1785. Late in 1788, when his intention of surveying the north line of the Ohio Company's purchase was thwarted by reports of the presence of Indian hunting parties, he lost faith in the ability of the American government to enforce its claims against

---

[1] A copy of Hutchins' first proposal made to the Board of Treasury in 1788 and quoted here in part, may be found in Papers Cont. Cong., CXXXVIII, Pt. I, 617. For his second proposal, addressed to Congress, see ibid., LXXVIII, Pt. XII, 541.

[2] Hutchins to Gardoqui, Pittsburgh, November 29, 1788, Facsimiles from Spanish Archives.

[3] See above, this section.   [4] See above, this section.

[5] Quattrocchi, op. cit., pp. 272-276. Survey of these small tracts, reserved for the use of Christian Delaware communities, was ordered in a resolution of September 3, 1788, although it had been already provided for in the Land Ordinance of 1785. The tracts were located north of the Ohio Company lands.

[6] Ibid., pp. 275-276. By the time Hutchins returned to Pittsburgh he had been obliged to send a messenger to New York for more money.

the tribes of the Northwest. Despite the fact that the Governor of the Northwest Territory, at this time, was organizing a conference with the aim of wresting new concessions from the Indians, Hutchins wrote to Gardoqui that he feared that the Indians would "forbid the settlement of their Country over the Ohio, or the sale of it by Congress."[1]

(4) Hutchins was deeply involved in the affairs of George Morgan, merchant and colonizer and old personal friend. For many years, these two men had been attentive to one another's interests.[2] During the year 1788, Morgan's attention had been diverted from a scheme for settling the Illinois country to a project for a colony on the western side of the Mississippi River, under Spanish rule.[3] Plans for this colony had been developed with the encouragement of Gardoqui, and Hutchins was fully apprised of the correspondence between the Spanish minister and Morgan. When Hutchins went so far as to finance Morgan's expedition from Pittsburgh down the Ohio River for the purpose of founding the colony, his defection from the United States was but a short step away. Hutchins, who obviously had great faith in Morgan's judgment and ability, hoped for deliverance from the vexations of his current employment by following the lead of his friend.[4]

Hutchins' reputation for devoted service to the cause of American westward expansion was saved by the termination of his life, which occurred at Pittsburgh before his offer to Spain was accepted. After an illness of several months, characterized by "a failing of the nerves and an almost insensible waste of the constitution," Hutchins died, April 28, 1789, at the age of fifty-nine years.[5]

No Geographer of the United States was appointed to succeed Hutchins,[6] but this did not prevent fulfillment of the

---

[1] Hutchins to Gardoqui, Pittsburgh, November 29, 1788, Facsimiles from Spanish Archives.

[2] For references to the friendship between Hutchins and Morgan, which accounted for Morgan's guardianship of Hutchins' son and Hutchins' service to Morgan as an advisor on western lands, see Quattrocchi, op. cit., pp. 261-263, 275, 280-281.

[3] Max Savelle, George Morgan, Colony Builder (New York: Columbia University Press, 1932), p. 203.

[4] Quattrocchi, op. cit., pp. 286-287. On Morgan's projected colony, New Madrid, see Savelle, op. cit., pp. 201-228.

[5] The Pennsylvania Gazette, May 2, 1789; quoted in Quattrocchi, op. cit., p. 295.

[6] Applicants for the office of Geographer, none of whom met with success, included William Morris, former surveyor for New York in the Seven Ranges, Andrew Ellicott and Andrew Porter, both of whom had participated in the survey of Pennsylvania's western boundary, and Rufus Putnam, later to become Surveyor General in the Northwest. (Applications for Office under President

federal obligation to bound the areas granted to the Ohio Company and to John Cleves Symmes and his associates. Completion of this work, and the surveying of townships and sections within these areas will comprise the subject matter of the following two sections of the present chapter.

Survey of the Ohio Company Lands

After Thomas Hutchins retired to Pittsburgh, there to end his days, Israel Ludlow and Absalom Martin continued the survey of boundaries encompassing the Ohio Company's purchase. While Martin proceeded up the Scioto River, Ludlow completed the charting of the Ohio River's course between the western limit of the Seven Ranges and the mouth of the Scioto.[1] After an interruption, during which the representatives of several Indian tribes submitted to American treaty terms, Ludlow resumed surveying.[2] He completed the enclosure of the Ohio Company's lands by running a line across the "top" of the area designated by Hutchins, between the Seven Ranges and the Scioto River (Fig. 14), and then surveying down the Scioto River to a point where Martin had left off. By the first week in June, 1789, Ludlow could report the fulfillment of the assignment which Hutchins had begun eight months before.[3]

These boundaries gave the Ohio Company more than enough room.[4] At first, the Company contracted for a tract (Fig. 14)

Washington, Manuscripts Division, Library of Congress.) Winthrop Sargent was prepared to leave his duties as Secretary of the Northwest Territory for the post of Geographer. (Pershing, "Winthrop Sargent," p. 140.)

[1] Quattrocchi, op. cit., p. 276.

[2] Ibid. The Treaties of Fort Harmar, signed January 9, 1789, during this interruption, are reproduced in Carter, Territorial Papers, II, 174-186.

[3] Quattrocchi, op. cit., p. 276; and Harmar to St. Clair, Fort Harmar, May 8, 1789, Northwest Territory Papers, Miscellaneous, 1787-1789, Manuscripts Division, Library of Congress. A manuscript map showing the completed boundaries, delivered to the Secretary of the Treasury September 17, 1790, may be found in Old Map File, Records of the General Land Office (Record Group 49), Cartographic Records Branch, the National Archives. It is titled, "A Survey of the Exterior Lines of That Tract of Country Sold to Messrs. Sargent & Cutler Containing Four Million, Nine Hundred and One Thousand, Four Hundred and Eighty Acres." Scale: 1 inch to 200 chains (2.5 miles).

[4] The boundaries were intended to accommodate--as is well known among students of this period of United States history--not

whose western boundary failed to come within fifteen miles of the
Scioto River, and whose northern boundary fell almost forty miles
short of the line which Ludlow ran between the Seven Ranges and
the Scioto, in 1789.[1] The Company's ability to pay did not permit the completion of purchase of even this amount of land, however. In 1792, a patent for somewhat more than half of this contractual area was issued to the Company, and a second patent was
granted for additional acreage covered by military warrants (Fig.
14).[2] Finally, Congress set aside a relatively small area, which
has come to be known as the Donation Tract (Fig. 14). It was held
in trust by the Company until it could be subdivided and distributed by lot among such pioneers as were willing to occupy positions exposed to Indian attack.[3]

With the Ohio Company's outer boundaries accounted for, we
may turn to a consideration of surveying within those borders--
surveying, undertaken by the Ohio Company itself, which extended
the effective life of the Land Ordinance of 1785. This internal
surveying began with the platting of the town of Marietta, at the
mouth of the Muskingum River, in the spring of 1788, and continued
at least into the year 1795.[4] In laying out townships, the Ohio

---

only the Ohio Company but also the "Scioto Associates," a group of
influential speculators. On this group, see Hulbert, Records of
the Ohio Company, I, lxx-xcii.

[1] The contract between the Company and the Board of Treasury
called for conveyance of title to this tract, totalling well over
1,500,000 acres, as soon as the Ohio Company had made payment of
$1,000,000 (ibid., pp. 31-32). A manuscript map showing the bound-
aries of the tract as surveyed by Israel Ludlow may be found in Old
Map File, Records of the General Land Office (Record Group 49),
Cartographic Records Branch, the National Archives. It is titled,
"A Survey of the Purchase Made by Winthrop Sargent & Manasseh Cut-
ler Esqrs. as Agents for the Ohio Company, Containing One Million
Seven Hundred & Eighty One Thousand Seven Hundred & Sixty Acres In-
cluding the Reservations." Scale: 1 inch to 200 chains (2.5 miles).
For evidence that Israel Ludlow surveyed the boundaries see Hulbert,
Records of the Ohio Company, II, 120, 139 and 230.

[2] For the law authorizing conveyance of both of these tracts
see ibid., I, cvx-cxviii.

[3] For authorization of this donation, see ibid. The legal
history of donated lands as well as other lands under the original
control of the Ohio Company is described in William E. Peters,
Ohio Lands and Their History (3d ed.; Athens, Ohio: W. E. Peters,
1930), pp. 166-179.

[4] An order to continue surveying, dated January 10, 1795,
appears in Hulbert, Records of the Ohio Company, II, 233.

Company's field workers simply extended westward the kind of surveying which had been begun in the Seven Ranges, but they went beyond this precedent by subdividing the townships, generally, into square-mile sections, and these in turn into lesser lots. At Marietta, and in various other parts of the Purchase, the surveyors staked out lots which bore no fixed relation to the grid decreed by the Land Ordinance, but such cases were considered exceptional, at the time of survey.[1]

Four men, by virtue of their participation in the Ohio Company surveys, bridged the gap between field work in the Seven Ranges and subsequent township surveying by the federal government. Three of these men, having engaged in the survey of the Seven Ranges, inaugurated the subdivision of the Ohio Company's lands,[2] and of this number one returned to federal surveying after passage of the Land Act of 1796.[3] A fourth man, Rufus Putnam, having supervised the Ohio Company's surveys, went on to assume general control of federal surveying under the Land Act of 1796.[4] Putnam, no stranger to the reader of earlier chapters in this study, rapidly established himself as the foremost leader of the Ohio Company colony. As Superintendent of Surveys for the Company, he was the first administrator of the contract system, a financial arrangement which he later brought over into federal public land surveying.[5]

The contract system, which Putnam, by the way, initially opposed, was adopted by the Ohio Company late in 1788, and further developed in resolutions of a later date.[6] It called for

---

[1] For expressions of official insistence upon adherence to the grid of the Land Ordinance of 1785, see ibid., pp. 92, 217. A manuscript map of the Ohio Company lands, showing watercourses projected on a strict Land Ordinance grid, may be found in Old Map File, Records of the General Land Office (Record Group 49), Cartographic Records Branch, the National Archives. It is titled "Plan of the Ohio Company Lands" and is signed by Rufus Putnam. Scale: 1 inch to 200 chains (2.5 miles).

[2] These three men were Ebenezer Sproat, Anselm Tupper and John Mathews. See "Contribution to the Opening of the Northwest," chap. ix, above.

[3] This was John Mathews.

[4] See "Federal Surveying Begins Again," chap. x, below.

[5] For notice of Putnam's appointment as Superintendent, see Hulbert, Records of the Ohio Company, I, 26 and II, 92, 138.

[6] Ibid., I, 68-70, 125, and II, 221, 231.

the engagement "of all persons alike, who may apply and appear to be competent to the business," for surveying at a set price per mile.[1] The system differed from another tried earlier by the Ohio Company, in assigning to the surveyor the status of an entrepreneur rather than that of a wage earner. It differed from the terms of survey which prevailed in the Seven Ranges, in requiring the surveyor to contract for the performance of a specified amount of work. It differed from both of its predecessors in throwing open the opportunity of subdividing land to any surveyor willing to risk the venture.

Of the surveyors with whom Putnam later signed contracts, under the Land Act of 1796, the most important was Israel Ludlow, already familiar as a surveyor in the Seven Ranges, and as the man mainly responsible for enclosing the Ohio Company's lands. We shall now be able to follow the activities of Ludlow up to the beginning of his period of service under Putnam, by turning to the subject of land subdivision in the Miami Purchase.

Survey of the Miami Purchase

During the years that the Ohio Company was gaining a foothold in the Northwest, a second tract on the Ohio was being developed, under the leadership of John Cleves Symmes. Symmes, who had applied to Congress for land soon after the Ohio Company's grant was obtained, operated as a lone proprietor backed by a group of New Jersey land speculators.[2] In this capacity, he received Congressional approval of a plan for purchasing a tract fronting on the Ohio, between the Great Miami and the Little Miami rivers (Fig. 14),[3] and late in the year 1788 set off from a base at Limestone (Kentucky) to explore and lay plans for the settlement of this territory, the "Miami Purchase."[4] In Symmes'

---

[1] Ibid., II, 221.

[2] For Symmes' application, August 29, 1787, see Carter, Territorial Papers, II, 70-71. On Symmes as proprietor, see Beverley W. Bond, Jr. (ed.), The Correspondence of John Cleves Symmes, Founder of the Miami Purchase (New York: The Macmillan Company, 1926), p. 4, n.

[3] Jrnls. Cont. Cong., XXXIII, 594. For contract prepared by the Board of Treasury, see Jacob Burnet, Notes on the Early Settlement of the North-Western Territory (New York: D. Appleton & Co., 1847), pp. 490-491.

[4] Symmes to Jonathan Dayton, Limestone, October 12, 1788, in Bond, Symmes Correspondence, pp. 44-47.

advance party was Israel Ludlow.[1]

From the beginning, Israel Ludlow served as Symmes' principal surveyor, and in due course he was additionally charged with the duties of a representative of the United States, relative to the Purchase. It fell to Ludlow, as a surveyor in the latter capacity, to determine the outer boundaries of the grant. Early in 1791, he was called upon by the Secretary of the Treasury to establish these boundaries, which he began to do, but a dispute over their definition rendered his work inconclusive.[2] By the spring of 1794, he was able to supply to the Secretary a chart showing three sides of the Purchase as finally determined: the courses of the two Miami rivers, and the course of the Ohio River between their mouths.[3] The fourth side, or northern boundary, of the Purchase was designated in a patent issued to Symmes later in the same year as "a parallel of Latitude to be run from the Great Miami River to the little Miami River so as to comprehend the quantity of three hundred & eleven thousand six hundred & eighty two acres of Land" (Fig. 14).[4] Ludlow did not mark this line on the ground,[5] but he had already surveyed a parallel of latitude

---

[1] Ibid. Both Symmes and Ludlow, having explored this territory in late September and early October, 1788, withdrew. Ludlow thereupon joined Hutchins for the Ohio Company boundary survey. During the interruption in that survey, which allowed for the completion of Indian negotiations at Fort Harmar, Ludlow returned to the Miami Purchase. On December 24, 1788, he shared in the founding of Cincinnati. (Symmes to Dayton, Northbend, May 18, 19 and 20, 1789, ibid., p. 60.) He departed from the Miami Purchase to resume the survey of the Ohio Company boundary in March, 1789. (Symmes to Dayton, Northbend, January 9, 1790, ibid., p. 115.)

[2] In keeping with Symmes' contract of 1788, Ludlow set about bounding the purchase on the east not by the Little Miami River, as in Fig. 14, but by a line originating at a point twenty miles up the Ohio River from the mouth of the Great Miami. (Field Notes of Symmes Purchase, Ohio, Case F, No. 99, Records of the General Land Office [Record Group 49], Cartographic Records Branch, the National Archives.) Symmes claimed as far east as the Little Miami, and in April, 1792, Congress acted to allow this claim. (Carter, Territorial Papers, II, 388-389.)

[3] See reference to this survey in report of Albert Gallatin to committee of House of Representatives (American State Papers, Public Lands, I, 76), and in final patent issued to Symmes (Carter, Territorial Papers, II, 496-498).

[4] Ibid. This line enclosed as much land as Symmes was able to pay for.

[5] Putnam to Secretary of the Treasury, Marietta, March 11, 1802, in Carter, Territorial Papers, III, 215.

further north, by way of connecting the two Miamis.[1]

As Symmes' principal surveyor, Ludlow supervised the subdivision of the Miami Purchase in keeping with the terms of the Land Ordinance of 1735.[2] For guidance in projecting township and section outlines throughout the Purchase, Ludlow, like Hutchins before him in the Seven Ranges, laid down a base line. The line was placed about as far south as the bends of the Ohio River would permit (Fig. 14), and from this base surveyors ran lines south to the River and north for a distance of about twenty miles.[3] Belatedly, when it was realized that section corners staked out independently along these north-south lines were not in agreement, Symmes directed Ludlow to establish a principal meridian, called "the standard," to which all east-west lines were to be adjusted (Fig. 14).[4] The Miami Purchase thus became the first field of survey to be divided into quadrants by a prime meridian and base line. The control exerted by these master lines was, however, negligible. Individual townships were numbered eastward from the Great Miami, without reference to the prime meridian, and ranges of townships were numbered northward from the Ohio, without reference to the base line.[5] Furthermore, boundaries in general were

---

[1] By drawing this line from the "head of the Little Miami" westward to the Great Miami, as part of his boundary survey submitted early in 1794, Ludlow was able to show that there was hardly more than five hundred thousand acres available between the two Miamis, rather than more than a million acres, as once had been supposed. (American State Papers, Public Lands, I, 76.) Survey notes on this line are included in Field Notes of Symmes Purchase, National Archives.

[2] Symmes, by his original contract of October 1787, was required to abide by the terms of a resolution of July 23, 1787, which had called for the subdivision of the Ohio Company lands "according to the land Ordinance of the 20th of May 1785." (Carter, Territorial Papers, II, 54-56.)

[3] Burnet, op. cit., pp. 418-419.

[4] Ibid. The principal meridian and the base line are shown on a map in Old Map File, Records of the General Land Office (Record Group 49), Cartographic Records Branch, the National Archives. The map is titled "Map of a Part of the Miami Country," and bears statement, dated 1814, that it is a copy of a map "in the possession of Mr. Thomas Henderson of Cincinnati." Scale: 1 inch to 2 miles. The principal meridian is shown passing two miles west of the eastern boundary of Twp. III, Fractional Range II, and the base line is shown paralleling the northern boundary of Fractional Range II, two miles south of it.

[5] Attention is called to this unusual system of township numbering in Sherman, Ohio Land Subdivisions, p. 69.

so freely run by compass that "scarcely two sections could be found in the Purchase of the same shape or of equal contents."[1]

Symmes, with full knowledge of the contrast between his own policy and that of the Ohio Company, allowed settlement to spread uncontrolled throughout his territory, and left most of the necessary surveying to be looked after by individual purchasers.[2] Operating accordingly, Ludlow engaged, with other men, in the lucrative activity of purchasing, laying out and developing various promising townsites, including that of Cincinnati, and he further advanced his interests by collecting fees for locating square-mile sections and lesser claims for numerous individuals.[3] In this business of locating claims, Ludlow and other surveyors functioned in much the same manner as federal surveyors would have done under the original Jefferson-Williamson plan of 1784. With warrants in hand for specified quantities of land, they independently selected and staked out lots in widely scattered locations, generally intending to render them conformable to the over-all grid.[4] The resulting lack of control, already referred to, demonstrated only too clearly what surveying would probably have been like throughout the public domain, had the Jefferson-Williamson plan been put into effect.

As the time approached when Congress would pass a new land law and turn away permanently from the policy of granting large tracts to private proprietors, Symmes unwisely began to sell land north of the limit of his patent, and Ludlow followed along northward with his surveying.[5] This ill-considered act of expansion

---

[1] Burnet, op. cit., p. 418. For a severe contemporary criticism of the quality of surveying within the Symmes Purchase, see St. Clair to Secretary of State, Cincinnati, July, 1799, in Smith, St. Clair Papers, II, 443-445.

[2] Symmes himself describes this contrast in Symmes to Dayton, Northbend, May 18, 19 and 20, 1789, in Bond, Symmes Correspondence, pp. 54-55.

[3] On Ludlow as a developer of townsites, see Henry B. Teetor, "Israel Ludlow and the Naming of Cincinnati," Magazine of Western History, II (July, 1885), 251-257. As an ambitiously active locater of claims Ludlow is cited in Symmes to Dayton, Northbend, January 9, 1790, in Bond, Symmes Correspondence, pp. 114-125.

[4] A "selling point" in Symmes' original approach to the public in the East was the opportunity held forth of purchasing land warrants for application to lots defined by a grid. (Burnet, op. cit., pp. 485-486.)

[5] Bond, Symmes Correspondence, pp. 16-21.

eventually led to the financial ruin of Symmes, and may have partially accounted for the rejection of an application which Ludlow later made for the Surveyor Generalship of the Northwest.[1] Eventually, the Ludlow surveys were officially extended northward under federal auspices, but the area of the grant to Symmes was never enlarged.[2]

---

[1] Timothy Pickering, as Secretary of State, cast suspicion upon Ludlow's motives in applying for the Surveyor Generalship in 1796 by pointing to his interest in "land affairs." (Pickering to President Washington, New York, July 19, 1796, Miscellaneous Letters, Department of State, Foreign Affairs Section, Legislative, Judicial and Diplomatic Records Branch, the National Archives.)

[2] The area of federal rectangular surveying north of the Symmes Purchase wherein the Ludlow surveys were continued appears today in combination with the Symmes Purchase as a distinct tract in which ranges of townships run east and west and sections are numbered like those in the Seven Ranges and the Ohio Company lands. See "Map of Ohio Showing Original Land Subdivisions," accompanying Sherman, Ohio Land Subdivisions. Scale: 1 inch to 6 miles.

CHAPTER X

HOW FEDERAL SURVEYING CAME TO BE RESUMED,
UNDER THE LAND ACT OF 1796

An event of paramount importance occurred during the very months when Congress was occupied with modifying its policy for the Northwest, in 1787. This was the framing of the United States Constitution, by a convention which met in Philadelphia from May to September of that year. By the summer of the following year a sufficient number of states had ratified the Constitution for it to go into effect, and in the spring of 1789 the government provided for by the Constitution began to function.[1] The new Congress, while attending to legislation deemed essential to the national welfare, during its first session, reenacted the Northwest Ordinance of 1787, but permitted the Land Ordinance of 1785 to expire.[2]

No new land law, to take effect under the Constitution, was passed by Congress until 1796. On three separate occasions, however, bills to meet the problem of western lands showed some promise of passage. On the first of these occasions, leadership in urging a land act was taken by Thomas Scott, Representative for the western counties of Pennsylvania, who wanted to reopen direct sales of land to men of small means, relieving them of the necessity of choosing between purchasing from land companies or settling illegally on the public domain.[3] Having spoken vigorously in favor of a system of indiscriminate locations modelled after the land policy of colonial Pennsylvania, Scott succeeded in inducing the House of Representatives to resolve that a bill for

---

[1] The Continental Congress, having declared that the new government would begin operations March 4, 1789, ceased to function as an organized body as of the middle of September, 1788. (Burnett, The Continental Congress, pp. 719-726.)

[2] The Northwest Ordinance was approved, with slight modifications adapting it to the Constitution, August 7, 1789. (Annals of Congress, 1st Cong., cols. 2159-2160.)

[3] Ibid., col. 629.

opening a land office "ought to pass." He was appointed chairman of a committee to prepare such a bill, and there the matter rested.[1] This occurred during the first session of the new Congress. During the second session, Congress turned to Alexander Hamilton, Secretary of the Treasury, for recommendations.[2] A bill passed by the House a year later embodied Hamilton's advice to the extent of providing for a General Land Office, subordinate land offices and a Surveyor General. Sale of land by contract, "within natural boundaries or lines or both," was allowed for, but the act was mainly concerned with the sale of land through warrants to be applied to lands indiscriminately located. The Senate postponed consideration of this bill, and again the matter rested.[3]

The evident tide running against re-adoption of the principles of the Land Ordinance of 1785 began to turn during the debate on the Secretary of the Treasury's report to Congress. Elias Boudinot of New Jersey, close associate of John Cleves Symmes in the Miami Purchase, rose to deplore the practice of indiscriminate location, and to remind the House that the Continental Congress "had adopted a method to obviate the inconveniences" of this mode of land disposal.[4] In speaking against indiscriminate location, he was joined by Hugh Williamson of North Carolina.[5] The presence of Williamson in Congress at this time is especially noteworthy. When for a third time the new Congress addressed itself to the problem of western lands, in 1792, it was Williamson who brought forward a land act, on behalf of a committee of which he was a member. This bill called for townships six miles square, like those described in the Land Ordinance of 1785. It was given a second reading in the House, and then neglected.[6]

Nearly four years elapsed after the dismissal of Williamson's bill, before both houses of Congress were ready to move with determination toward the passage of a land act. To provide a necessary background for the deliberations of Congress at this

---

[1] Ibid., cols. 628-629, 665-666.   [2] Ibid., col. 1072.

[3] Ibid., cols. 1840-1842, 1866, 1964, 1973-1974. Hamilton's report to Congress appears in American State Papers, Public Lands, I, 8-9.

[4] Annals of Congress, 1st Cong., col. 1831.   [5] Ibid.

[6] Ibid., 2d Cong., cols. 573, 574. The content of the bill is inferred from later reference to it, ibid., 4th Cong., 1st Sess., col. 331.

later date, two brief discussions follow. In the first, an account is given of events leading up to a crucial victory over the Indians of the Ohio country--a victory which the United States had little reason to hope for, when Williamson's bill was before Congress. In the second, the reader is prepared for the fact that when Congress finally took action, in 1796, land companies received no consideration whatever. Following these two sections, the passage of the Land Act of 1796 is discussed, and the chapter closes with an account of the renewal of U.S. public land surveying.

The Indians Sustain a Decisive Defeat

When organized settlement of the Northwest Territory began, beyond the Seven Ranges, the advancing Americans hoped to avoid an Indian war. Arthur St. Clair, governor of the Territory, arranged for peace negotiations with the Indians as one of his first official acts,[1] and both Symmes and the Ohio Company initially adopted a conciliatory attitude toward the natives.[2] St. Clair soon discovered, however, that the Indians were reluctant to treat with him.[3] Symmes began to realize that the Shawnee, in whose territory he had settled, were not prepared to retire peaceably, and the Ohio Company found itself confronted by Delaware and Wyandot who were persistently hostile.[4] As time went on, resistance increased, and impatience with the Indians became general among Americans in the Northwest.

To put an end to a series of thefts and small raids, Governor St. Clair decided on war in June, 1790, directing his attention toward the upper Wabash and Miami river valleys, home of the Miami tribe and center of Indian resistance. An expedition sent

---

[1] St. Clair to Secretary of War, Philadelphia, January 27, 1788, in Smith, St. Clair Papers, II, 40-42.

[2] For evidence of this attitude see General Parsons to Mrs. Parsons, Muskingum River, June 1, 1788, in Hall, Life of Parsons, p. 521, and Symmes to Dayton, Limestone, October 12, 1788, in Bond, Symmes Correspondence, p. 47.

[3] St. Clair to Secretary of Congress, Fort Harmar, November 6, 1788, in Smith, St. Clair Papers, II, 97. Few of the principal chiefs of the tribes represented signed the Treaties of Fort Harmar, January 9, 1789. (Downes, Council Fires, p. 305.)

[4] Symmes to Dayton, Lexington, April 30, 1790, in Bond, Symmes Correspondence, pp. 127-128, and Putnam to President Washington, New York, July 24, 1790, in Buell, Memoirs of Rufus Putnam, pp. 232-233.

into this territory a few months later suffered heavy losses, and a second expedition which St. Clair himself led northward from Cincinnati, in the fall of 1791, was utterly defeated. While hope rose among the Indians that they would win back all lands north of the Ohio, the Americans laid plans for a third expedition into the country of the Miami. General Anthony Wayne, new commander of the Army in the West, received authority to renew hostilities in September, 1793, but deferred his invasion to allow for thorough preparation. After wintering at Fort Greenville, sixty miles north of Cincinnati (Fig. 15), Wayne's army was joined by one thousand mounted Kentucky volunteers, and the combined forces opened a skillfully executed campaign of conquest.[1]

Wayne attained victory at the Battle of Fallen Timbers, in August, 1794, after having occupied strategic points in the homeland of the Miami, including the present site of Fort Wayne, Indiana (Fig. 15). Refraining from immediate dictation of peace terms, he returned to his headquarters at Fort Greenville, there to cultivate dissension among the allied Indian tribes. In this, he was aided above all by a deep disappointment which the tribes felt toward the British at Detroit, who had urged the Indians to meet Wayne in battle, only to fail them when they most needed supplies and reinforcements. What neither Wayne nor the commandant at Detroit knew at the time of Wayne's campaign was that negotiations in London were leading toward the complete removal of the British from posts in the Northwest. When Wayne convened a treaty conference at Fort Greenville in the summer of 1795 he was able to confront the tribesmen with news that, under the terms of Jay's Treaty, signed in London in October, 1794, the British were soon to surrender to the Americans those strongholds toward which the Indians had looked for support for more than a decade while resisting American aggression. Indian capitulation was now at hand.[2]

By the Treaty of Greenville, dated August 3, 1795, four tribes--the Delaware, Wyandot, Chippewa and Ottawa--submitted to essentially the same terms of cession as those to which they had

---

[1] On the provocation and the organization of expeditions against the Indians, 1790-1794, see Downes, Council Fires, pp. 310-327.

[2] This account of the Greenville negotiations is based upon Dwight Smith, "Indian Land Cessions in the Old Northwest, 1795-1809" (Unpublished Ph.D. dissertation, Dept. of History, Indiana University, 1949), pp. 83-98.

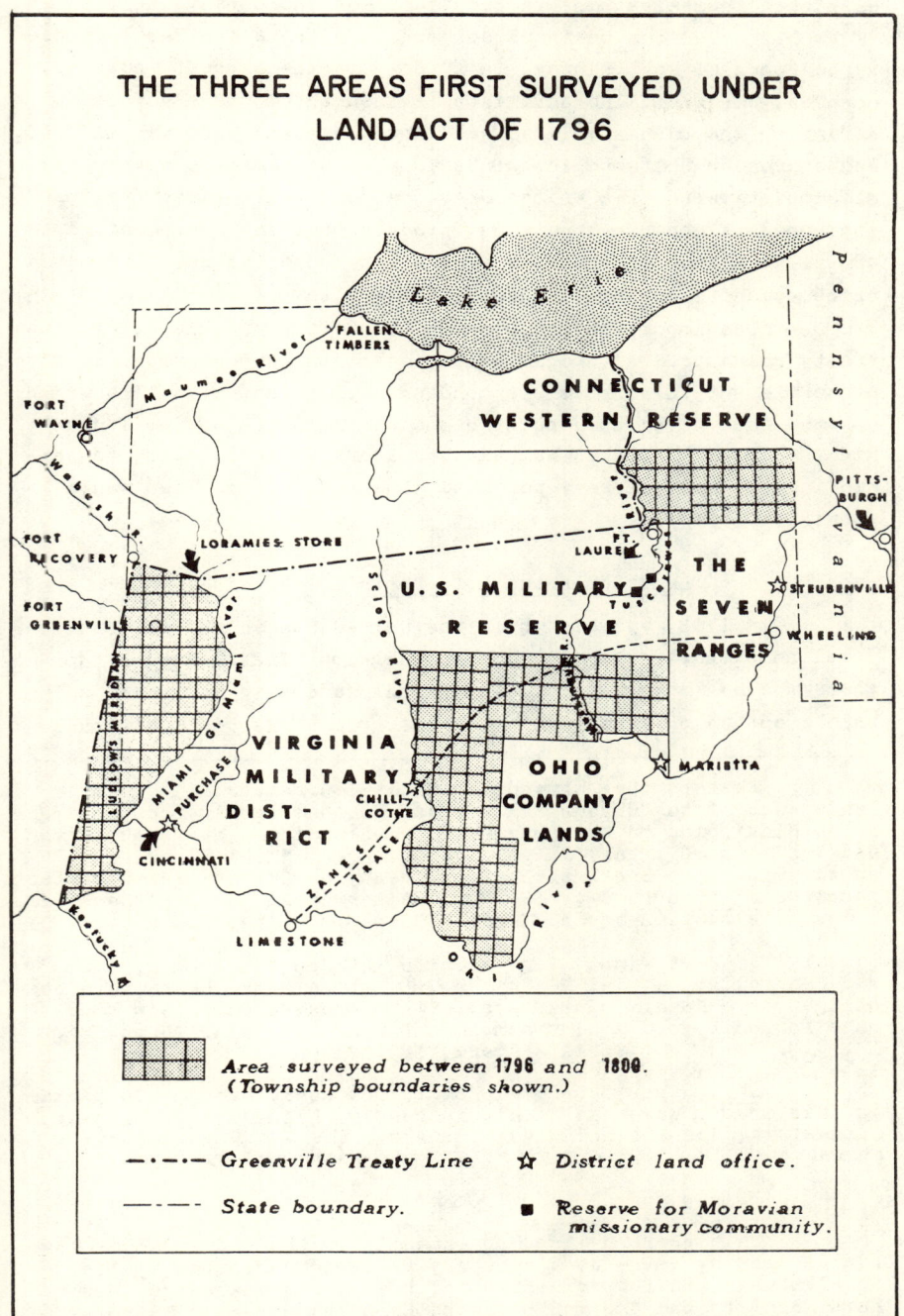

Fig. 15

committed themselves in the Treaty of Fort McIntosh, in 1785 (Fig. 3).[1] The Shawnee were subjected to terms similar to those agreed upon at Fort Finney, in 1786.[2] In the present instance, however, the Miami and affiliated tribes added their voices in affirming the boundary lines laid down between white man and red, and surrendered strategic points located as far west as the Mississippi River.[3] The United States accepted all of the tribes concerned as dependent nations, pledging the establishment of an effective system of trade, and offering to assist any interested tribe toward adoption of the white man's ways.[4]

Most important from the point of view of this study, the treaty guaranteed with a certainty hitherto unknown the security of settlement in territory extending as far west as a line which met the Ohio River at a point opposite the mouth of the Kentucky River (Fig. 15).[5] This was the territory to which, with the exception of lands already patented, the Land Act of 1796 would specifically apply.

Land Companies Lose Their Leadership

In 1796, by which time the United States had at its disposal an extensive territory secure against Indian attack, Congress was no longer willing to assign leadership in survey and settlement to private land companies. This change in attitude

---

[1] These tribes agreed, again, to surrender all claim to lands east of the Cuyahoga River and south of a line connecting a crossing point on the Tuscarawas branch of the Muskingum River and the site of a trading post on the Great Miami River, as shown, Fig. 15. See Article 3 of Treaty of Greenville in Carter, Territorial Papers, II, 526. On earlier cession along these same lines, see section headed "Indian Cessions," chap. i, above.

[2] At Fort Finney (see sections headed "Surveying, 1786," and "Surveying, 1787-1788," chap. vi, above) the Shawnee had been obliged to cede all claims east of the Great Miami. Now they were forced beyond a line somewhat further west, as shown, Fig. 15. (Carter, Territorial Papers, II, 526.)

[3] Ibid., pp. 526-527. The tribes newly brought to terms in this treaty were the Miami, Kaskaskia, Kickapoo, Piankaskaw, Potawatomi, Wea and Eel River Indians. The strategic points surrendered included the sites of Fort Wayne and Chicago.

[4] Ibid., pp. 529-530.

[5] This western line, objected to by the Indians, was insisted upon by Wayne as a boundary rendered desirable by its straightness and its connection of the two well known points of Fort Recovery and the mouth of the Kentucky River. (Smith, "Indian Land Cessions," pp. 96-97.)

since 1787, when the Continental Congress had been ready to consign the future of the Northwest to land companies, was due in part at least to disappointments which had been experienced in the meantime. Neither the Ohio Company nor the Symmes group had been able to fulfill the original terms of its contract,[1] and of three speculative associations which had applied for large grants in the Illinois country, none had been found willing to conclude a contract on terms acceptable to the government.[2]

Perhaps land companies would still have enjoyed a position of favoritism, had it not been for a taint of scandal which had become associated with their operations. The principals of the Ohio Company had gotten deeply involved in a secret connection with certain influential persons who cloaked their speculative ambitions with the reputation of the Company for responsible colonizing intentions.[3] Symmes aroused the distrust of Congress by persisting in selling land beyond the limits of his acknowledged claims.[4] Suspicion toward land companies in general was increased by the so-called Yazoo Companies, which challenged federal prerogatives south of the Ohio River.[5] Above all, large-scale land acquisition

---

[1] The Ohio Company failed to "fill out" the boundaries of the 1,500,000 acre tract defined in its contract of 1787, it will be recalled. The patent to Symmes, issued in 1794, covered only a fraction of the quantity of land contemplated in his contract of 1788. The patent was issued as result of a plea for "so much land as had already been paid for." (American State Papers, Public Lands, I, 76.)

[2] The petition of one of these associations, headed by Royal Flint, for lands on the Wabash and Mississippi Rivers (Carter, Territorial Papers, II, 74-76), precipitated the general million-acre invitation issued by Congress in October, 1787. This was followed by an application for large grants on the same two rivers by the Illinois and Wabash Company (ibid., pp. 115-117), and by a petition from George Morgan for a purchase on the Mississippi (ibid., pp. 112-115). Morgan, as we have seen, presently shifted his attention to colonization on Spanish lands west of the Mississippi.

[3] Hulbert, Records of the Ohio Company, I, lxx-xcii.

[4] Bond, Symmes Correspondence, pp. 17-18.

[5] In particular, these companies, with investments in Georgia's western claims, gave offense by undertaking directly the extinguishment of Indian titles. The companies became a subject of national scandal in February, 1796, when Georgia voided its earlier sale of western lands because of alleged bribery of members of the granting legislature. For a studied summing-up on the Yazoo Companies see Shaw Livermore, Early American Land Companies: Their Influence on Corporate Development (New York: The Commonwealth Fund, 1939), pp. 146-162.

was brought into disrepute in Congress by charges of attempted bribery which were aired before the House of Representatives early in 1796.

In an unusual hearing held by the House of Representatives in January, 1796, it was established that one Robert Randall, having visited Detroit where he entered into an association with Canadian merchants and others for the purpose of land speculation, had come to Philadelphia and canvassed members of both houses of Congress with a view to gaining their approval of a plan for acquiring title to the entire lower peninsula of the present state of Michigan. To improve the prospects of the scheme, Randall and a companion named Charles Whitney had offered shares of stock to the legislators. At the end of the hearing, Whitney was allowed a discharge on technical grounds and Randall was simply reprimanded, but the effect of the indiscretion of these two men lived on after their release.[1]

Once Randall and Whitney had been denounced by Congressmen whom they had approached, and the House had made a public show of parliamentary dignity outraged, the subject of large-scale land speculation became unwelcome in Congress. The prospect for further encouragement of private land companies in the Northwest had disappeared.

Passage of the Land Act of 1796

After passage by Congress, the Land Act of 1796, entitled "An Act providing for the Sale of the lands of the United States, in the territory north-west of the river Ohio," was approved by President George Washington on May 18, 1796.[2] The only immediate effect of the law was the reopening of sales of land in the Seven Ranges.[3] Surveying authorized by the law was confined to three

---

[1] Annals of Congress, 4th Cong., 1st Sess., cols. 200-229, 237, 244. Of unusual interest are the basic parliamentary issues which were raised in the House prior to the opening of this hearing (ibid., cols. 166-170, 171-183, 184, 185-195).

[2] The law is reproduced in Carter, Territorial Papers, II, 552-557.

[3] For authorization, see ibid., p. 555. Under this authorization a single quarter-township had been sold at Philadelphia and less than the equivalent of two townships at Pittsburgh, by June, 1798. (Report of Gallatin to House of Representatives, American State Papers, Public Lands, I, 82.) These were the first sales to take place since the auction of land in the Seven Ranges, in 1787.

areas, north of the Seven Ranges, west of the Seven Ranges, and west of the Symmes Purchase, respectively. Despite these limitations of reference, the act became the legal basis of all subsequent U.S. public land surveying.

That the Land Act of 1796 reestablished the basic principles of the Land Ordinance of 1785 is a fact which anyone may confirm by a comparative reading of the two laws. Yet the debates of Congress, in 1796, were almost completely devoid of comment on the land act of nine years before. The influence of the earlier law was apparently exerted through the Williamson's proposed bill of 1792, and through the land policy of New York State, which was indebted to the Land Ordinance of 1785.[1] If there was substantial agreement between the two federal land laws, there were also differences which, in view of the apparent lack of direct connection between them, should not occasion surprise. Both the similarities and differences, in so far as they relate to surveying, will be discussed in the following paragraphs.

Design of the surveyed grid. The Land Act of 1796 directed that, prior to sale at public auction, the land should be divided "by north and south lines . . . and by others crossing them at right angles."[2] Here was a borrowing from the Land Ordinance of 1785, expressed in nearly the terms of the original; yet, for a demonstration of the workableness of this system, the attention of Congress was directed in the course of debate not to the Ohio country, where the Ordinance had been applied, but to New York State, where such "ideal lines" were said by Representative Havens to have caused "little embarrassment."[3] The idea of rectangular surveying was apparently taken from Williamson's proposed bill of 1792,[4] and was adopted by the committee which drew up the Land Act of 1796 "so as to make titles certain."[5] Security of land title, in turn, was considered necessary "to get the highest price for the land."[6] Interestingly, the principle of prior survey met with no objection. Advocates of indiscriminate location were not heard from. But the laying out of surveys in a strict rectilinear

---

[1] This indebtedness of New York's system to national land law has been pointed out above, p. 61.

[2] Carter, *Territorial Papers*, II, 553.

[3] *Annals of Congress*, 4th Cong., 1st Sess., col. 336.

[4] *Ibid.*, col. 331.   [5] *Ibid.*, col. 330.   [6] *Ibid.*

grid was approved only over the protest of members who wished to take natural boundaries into account, as had Rufus King, in 1786.[1]

Like the Land Ordinance of 1785, the Land Act of 1796 required the rectangular lines to be run "so as to form townships of six miles square."[2] This spacing of the lines, at first rejected as found in Williamson's proposed bill by the committee which brought out the Land Act,[3] was restored by Congress, apparently on the instance of Havens and Van Allen from New York State.[4] Anything so venturesome as Jefferson's measurement in geographical miles, of 1784, was not even considered, but his idea of ten-mile spacing reappeared in debate, as it had done in Hamilton's report of 1790.[5] Also brought up in debate was a plan which offered some of the convenience of Jefferson's original scheme. It called for five-mile townships which were to be quartered into parcels of four thousand acres, these to be quartered again into thousand-acre tracts.[6] Rejected at this time, the idea was adopted shortly afterward for application in that military reserve adjoining the Seven Ranges which the Continental Congress had set aside in vain, and which the new Congress now covered in a comparion-law to the Land Act of 1796.[7] The five-mile township was also adopted for Connecticut's Western Reserve, and then fell into total neglect.[8]

Township and section. As in the Land Ordinance of 1785, alternate townships were ordered to be sold in square-mile sec-

---

[1] Ibid., cols. 329, 336.

[2] Carter, Territorial Papers, II, 553.

[3] Annals of Congress, 4th Cong., 1st Sess., col. 331.

[4] Ibid., cols. 330, 339.

[5] Ibid., cols. 334-335. For ten-mile suggestion in Hamilton's report, see American State Papers, Public Lands, I, 8.

[6] Annals of Congress, 4th Cong., 1st Sess., col. 343.

[7] For substitution of five-mile for six-mile townships in this companion law, see ibid., col. 1384.

[8] If Sherman, in his Ohio Land Subdivisions, p. 84, is correct in dating adoption of the five-mile township for the Western Reserve, then the chronology of the five-mile township idea is as follows: (1) suggested for national lands by Representative Kittera of Pennsylvania, February 17, 1796; (2) adopted by Connecticut Land Company, April 12, 1796; (3) adopted by House for U.S. Military reserve, May 17, 1796.

tions.[1] This alternation was expressive of a compromise, as it had been in 1785. But anyone reading the debates of 1796 is forced to recognize, along with an earlier writer, that the questions which divided Congress in 1785 had lost their importance.[2] Townships were no longer thought of as "holding forth an inducement for neighborhoods . . . to confederate for the purpose of purchasing and settling together,"[3] and purchase by sections was no longer fought for as a means of approximating that freedom of choice which characterized indiscriminate locations. Townships were simply spoken of in the debates as "tracts" or "parcels," which were expected to prove attractive to speculators either as wholes or in quarters. Sale by section[4] was looked upon as a means of bringing land closer to the financial reach of actual settlers.[5]

Conflict in Congress over the Land Act of 1796 arrayed those members whose first concern was revenue against those who were interested in accommodating settlers in the Northwest. Members of the former category were generally from the eastern seaboard, and were of greater influence in the Senate than in the House. They favored a relatively high minimum price per acre, and were disposed to give preference to large purchasers. The democratic complexion of the House was most clearly shown by approval of a clause--which the Senate struck out--providing for the sale of one-fourth of the land in quantities of less than a square mile.[6] Combatting the contention that this would unreasonably increase the cost of surveying, a group led by Repre-

---

[1] Carter, Territorial Papers, II, 553. Not only were square-mile sections to be sold; they were now to be partially surveyed.

[2] This change is noted in Treat, National Land System, p. 87.

[3] Grayson to Washington, April 15, 1785, in Burnett, Letters of Members, VIII, 95.

[4] "Section," as a term referring to a square-mile unit of land subdivision, reappeared in the Land Act of 1796, after having been dropped from the Land Ordinance of 1785 (see pp. 56-57, above). For this use of the term by John Cleves Symmes, see Symmes to Dayton, Northbend, May 18, 19 and 20, 1789, in Bond, Symmes Correspondence, p. 68.

[5] For expressions of the changed attitudes of 1796, see record of debate in Annals of Congress, 4th Cong., 1st Sess., cols. 328-331, 334-337, 338-344, 345-349, 350-355, 400-401, 402-423, 856-867.

[6] Ibid., col. 865.

sentative Crabb proposed that, since the land was to be sold at auction, the resulting increase of competition would raise the price of the land.[1] This group, whose real aim was the offering of tracts suitable to the pockets of the poor, succeeded in securing approval of one-hundred-sixty-acre lots; and Crabb pointed out that "it might require two or more poor men" to purchase even this relatively small amount at the agreed-upon minimum price of two dollars per acre.[2] Crabb was ahead of his time.

<u>Surveying and numbering</u>. In setting forth surveying regulations, the legislators of 1796 introduced certain innovations. To the stipulation of 1785 that all lines should be plainly marked upon trees, they added a specific requirement that trees should be marked near each section corner, one within each section for which a corner might stand.[3] To the requirement that a chain be used for measurement, they added a qualification designating the exact kind of chain to be used.[4] While calling for the preparation of plats of townships, they recognized field books as the primary evidence upon which plats must be based, and required submission of these books by surveyors.[5] Finally, in their concern for making land titles secure, they provided, as the framers of the Land Ordinance of 1785 had not done, for subdivision on the ground of those townships which were to be offered for sale by square-mile section. True, the Senate, in its desire for economy, limited subdivision to lines to be run at intervals of two miles, thus allowing for the marking of only three corners of each section (Fig. 16A), but this was a long step forward from the lines-on-paper of the Seven Ranges.[6]

Of particular interest are the few words in the Land Act which required that all lines be run "according to the true me-

---

[1] Ibid., cols. 858-865.   [2] Ibid., col. 860.

[3] Carter, <u>Territorial Papers</u>, II, 553.

[4] "All lines shall be . . . measured with chains containing two perches of sixteen feet and one half each, subdivided into twenty five equal links." (<u>Ibid</u>., p. 554.) This, of course, was a Gunter's chain, of half the full sixty-six foot length. See pp.

[5] Carter, <u>Territorial Papers</u>, II, 554.

[6] <u>Ibid</u>., p. 553. For record of Senate's amendment calling for an interval of two miles rather than one mile between lines, see <u>Annals of Congress</u>, 4th Cong., 1st Sess., col. 83.

ridian."[1] Hereby, an essential provision of the Land Act of 1785, which had been repealed for the convenience of surveyors in the Seven Ranges, was restored. There is no indication, however, that Congress thought of itself as acting to correct a mistaken earlier concession to surveyors. We have here, to all appearance, independent legislation originating with a representative from western Pennsylvania who recognized, as Thomas Jefferson had done a dozen years earlier, that this method of surveying "would be of essential service."[2] Acting again with the intention of making land titles secure, Congress ignored the complaint of a representative from New York that this restriction would make the survey "liable to many difficulties."[3] Within a few years, an appeal for release from the restriction would be heard from the field.[4] Fortunately, for the sake of precedent, it too would be ignored.

While specifying a definite order for the numbering of sections within townships--"beginning with number one, in the north east section, and proceeding west and east alternately" (Fig. 16B) --the Land Act merely indicated an expectation that townships themselves would be identified by numbers, and left it at that.[5] The adoption of a new order of section numbering combines with other evidence strongly to suggest that the legislators of 1796 were unaware of practices which had grown up under the Land Ordinance of 1785. Under the Ordinance, as will be recalled, the Board of Treasury had applied a numbering system (Fig. 10B) which would have served future surveying as well as, or better than, the system prescribed in the Land Act.[6] As to the numbering of townships, it is interesting to note that the generally admired system of numbering which has come to characterize federal public land subdivision throughout the United States did not originate

---

[1] Carter, Territorial Papers, II, 553.

[2] These are the words of Representative Kittera (Annals of Congress, 4th Cong., 1st Sess., col. 422). On Jefferson's original proposal of this restriction on surveying, see p. 77, above.

[3] Objection by Representative Van Allen, Annals of Congress, 4th Cong., 1st Sess., col. 422.

[4] Rufus Putnam to Congress, Marietta, March 10, 1786, in American State Papers, Public Lands, I, 83.

[5] Carter, Territorial Papers, II, 553.

[6] See Fig. 10B. This earlier order of numbering corresponded to the order of section-surveying eventually adopted in the field.

in the Land Act, as is often supposed. This system, which will be described in the next chapter, was instituted in the course of administering the Land Act, several years after its passage.

Surveyor General and deputy surveyors. For looking after the subdivision of the land, there was to be a Surveyor General and "skilful surveyors" who were to serve as his deputies.[1] A Surveyor General had first been proposed for federal surveying in 1790, when Hamilton's report called for such an officer to take up the duties of the late Geographer.[2] This is essentially what the Surveyor General in the Land Act was to do. Deputy surveyors in the Land Act were expected to function in the field in much the same manner as the surveyors of the Seven Ranges, and their duties with respect to description of the land were set forth in words nearly the same as those of the Land Ordinance of 1785.[3] They were not, however, expected to be "gentlemen surveyors," each officially representing a state, as were the men to whom the Ordinance referred, nor were their terms of employment the same as those stated in the earlier law. These terms will now be discussed.

It is easy to suppose that the contract system for the employment of deputy surveyors originated in the Land Act of 1796, since it began to operate soon after passage of the law. Actually, this system, which placed responsibility upon the deputy surveyor for "the wages of chain carriers, markers and every other expence [of surveying],"[4] was complete under the Land Ordinance of 1785, save for the essential feature of a contract binding the surveyor to the performance of a definite surveying assignment. The Land Act of 1796 not only failed to require contracts but neglected to specify that deputy surveyors should assume the responsibilities enumerated above. The Act simply provided that "the President of the United States may fix the compensation of the Assistant [that is, deputy] surveyors, chain carriers and axemen," adding that the total outlay for surveying should not exceed a certain number of dollars per mile.[5] Within these terms, the contract system was soon established by Rufus Putnam, as will

---

[1] Carter, Territorial Papers, II, 552-553.

[2] American State Papers, Public Lands, I, 9.

[3] Carter, Territorial Papers, II, 554.   [4] Ibid., p. 13.

[5] Ibid., p. 557. Payments were not to exceed three dollars per mile.

be described in the next chapter.

In assigning work to his deputy surveyors, the Surveyor General was to pay first attention to the demarcation of the Greenville Treaty Line.[1] At last, that "boundary line of property for separating the settlements of the citizens from the Indian villages and hunting grounds,"[2] envisioned in 1784, was to become a reality. By this late date, the amount of land eligible for federal surveying on the white man's side of the line had been reduced not only by Virginia's large claim along the Ohio River and Connecticut's Western Reserve along the shore of Lake Erie, but also by the Seven Ranges, the Ohio Company's territory and the Symmes Purchase. In a companion-law to the Land Act of 1796, already mentioned, Congress further reduced the area to which the Land Act would apply, by setting aside a tract for the satisfaction of military warrants (Fig. 15).[3] Under the Land Act, then, deputy surveyors were to attend to the subdivision of land in these locations:

(1) North of the Geographer's Line. Here, where Thomas Hutchins and his surveyors had originally intended to extend their work, unsurveyed territory stretched northward to Connecticut's Western Reserve (Fig. 15).

(2) West of the Seven Ranges. This was a field of survey limited to hardly more than twice the size of the first-named area, owing to the interposition of three large grants (Fig. 15).

(3) West of the Symmes Purchase. Here, surveying was placed upon the threshold of that great westward advance which would eventually carry it to the shores of the Pacific Ocean. For the moment, prospective surveying in this quarter was confined by the Greenville Treaty Line (Fig. 15). The line, however, would prove to be but the first of many boundaries, all solemnly agreed upon, which would yield to renewed American demands for land.

---

[1] Ibid., p. 553.   [2] Jrnls. Cont. Cong., XXV, 686.

[3] This law, approved June 1, 1796, not only set aside a large tract for the location of military bounty land claims (Fig. 15), but also provided for the survey of three small reserves within that tract, each containing a community founded by Moravian missionaries to the Delaware Indians. (Annals of Congress, 4th Cong., 1st Sess., cols. 2937-2938.) The law brought to an end a period of experimentation by Congress with the bounty land problem, writes W. T. Hutchinson, in his "Bounty Lands of the American Revolution," p. 101. By this time, of an estimated total of 1,850,800 acres in bounty claims, the Ohio Company and the Symmes enterprise had brought about the cancellation of 238,150 acres (ibid., p. 95). A few soldiers apparently had located their warrants in the Second Range of the Seven Ranges (ibid., p. 98, n.), and Congress had occasionally allowed an individual claimant to locate elsewhere, by specific resolution (ibid., p. 100).

Federal Surveying Begins Again

The first official act to take place under the new land law was the appointment of a Surveyor General. President Washington chose Simeon De Witt, who, as a member of Washington's headquarters during the Revolutionary War, had shared with Thomas Hutchins the title, "Geographer to the United States."[1] De Witt, however, "after weighing all circumstances," declined the appointment in favor of continuing as Surveyor General of New York State.[2] If the United States thus lost a distinguished surveyor, it gained in his stead an acknowledged leader in the Ohio country. This was Rufus Putnam, who was appointed to the office of Surveyor General, October 1, 1796.[3]

Putnam, having supervised the survey of lots for French settlers at Gallipolis on behalf of the federal government, in 1795,[4] and having been chosen to superintend the survey of Zane's Trace, in June, 1796,[5] looked upon the Surveyor Generalship as "the Last & best gift I recived from President Washington."[6] The fact that the President selected him only after several other candidates had been considered, in addition to De Witt, was unknown to Putnam.[7] Putnam served until 1803, when a man techni-

---

[1] De Witt's departure from this office, in favor of the post of Surveyor General of New York State, has been noted, p. 70, above.

[2] De Witt to Timothy Pickering, Philadelphia, June 25, 1796, Miscellaneous Letters, Dept. of State, May-July, 1796, Foreign Affairs Section, Legislative, Judicial and Diplomatic Records Branch, the National Archives.

[3] Pickering to Putnam, Department of State, October 1, 1796, in Buell, Memoirs of Putnam, p. 412. Putnam, the leading spirit in the conduct of the Ohio Company's affairs, was also one of three judges of the Northwest Territory, at this time.

[4] Putnam to Oliver Wolcott, Marietta, May 12, 1796, ibid., pp. 409-410.

[5] Ibid., p. 125. Zane's Trace, the only important road in the Ohio country up to 1800 (Fig. 15), was opened through the forest in 1796 by Ebenezer Zane, last encountered in this study as a pioneer settler at Wheeling who assisted the surveyors of the Seven Ranges.

[6] Ibid.

[7] Three men who applied for the office by letter were Elijah Backus, Israel Ludlow and Joseph Neville. (Applications for Office under President Washington, George Washington Papers, Manuscripts Division, Library of Congress.) Two men who were

cally more competent was appointed by President Jefferson to supplant him.[1] Whatever his technical shortcomings may have been, and however bitter he may have been over a displacement which he regarded as purely political in motivation,[2] Putnam deserved the honor that was his, of following Hutchins in managing the survey of federal lands. Many years earlier, his had been among the strongest voices championing the settlement of the Ohio country, and he had never relinquished his devotion to that end. Furthermore, as former Superintendent of Surveys for the Ohio Company, he represented a direct link with surveying under the Land Ordinance of 1785.

To complete the proper scope of the present study, we need proceed no further than the first three seasons of surveying under Putnam, in the years 1797, 1798 and 1799.[3] At the conclusion of this period, which was equivalent in length to the time required for the earlier survey of the Seven Ranges, field work assigned in the Land Act of 1796 was practically complete. The beginning --the second beginning--of U.S. public land surveying had come to an end. Throughout the period, Putnam placed principal reliance upon Israel Ludlow, who, as a veteran of the Seven Ranges and mainstay of land subdivision in the Symmes Purchase, provided a second link with surveying under the Land Ordinance of 1785.

At Putnam's request, Ludlow undertook the running of the Greenville Treaty Line (Fig. 15), beginning at what was called

---

considered for the office in preference to Putnam were Lieutenant Governor Wood of Virginia, and a Major Alexander, also from Virginia. (Pickering to President Washington, September 29, 1796, Miscellaneous Letters, Dept. of State, August-December, 1796.)

[1] Putnam's successor, Jared Mansfield, left his post as professor of mathematics at the United States Military Academy to become Surveyor General. He had recently published a book, Essays, Mathematical and Physical . . . (New Haven, Connecticut [1802]), and was soon to demonstrate his professional skill by conducting observations of a comet which appeared in 1807, and calculating its orbit.

[2] Putnam believed that his removal was due solely to Jefferson's hatred of Federalists, of whom he was one. (Buell, Memoirs of Putnam, pp. 125-126.)

[3] A surveying season, said Putnam, lasted from "the opening of the spring untill neerly the setting in of Winter." (Putnam to Secretary of the Treasury, Marietta, October 7, 1797, Letters Received from the Surveyor General, Northwest Territory, Records of the General Land Office [Record Group 49], Interior Section, Natural Resources Records Branch, the National Archives.)

in the Treaty "the crossing place above Fort Lawrence."[1] During the survey season of 1797, Ludlow ran the line southwesterly a distance one hundred and fifty miles to a point on the upper waters of the Great Miami River,[2] while other men began the survey of the tract which Congress had set aside for the satisfaction of military warrants.[3] By the end of the survey season of 1798 Ludlow had carried the Treaty Line no further, but in the meantime he had not only surveyed a large part of the military tract, but had begun to subdivide that remote area of public land which lay west of the Symmes Purchase.[4] Ludlow completed the Treaty Line in 1799, carrying it first to Fort Recovery, near the present Ohio-Indiana border (Fig. 15), and then southwesterly to a point on the Ohio River opposite the mouth of the Kentucky River.[5] The Indians had wanted a great swathe cut through the forest all along the line, "that it might prevent the White people from settling on their hunting grounds," but the ever present motive of economy prevented the hire of axe-men for the job.[6] Throughout its length, the course of the boundary was identified "like any other line of Survey in the woods," that is, by stakes and marked trees.[7]

---

[1] Putnam travelled north to personally mark the beginning point. (Putnam to Secretary of the Treasury, Marietta, July 22, 1797, ibid.) This point, at a crossing place over the Tuscarawas branch of the Muskingum River, was at "Tuscarawas" (on the site of modern Bolivar), as previously mentioned. It was hardly more than a mile upstream from Fort Laurens (Fig. 15).

[2] Putnam to Secretary of the Treasury, Marietta, May 18, 1797 and September 2, 1797, ibid. The point on the upper waters of the Great Miami was at Loramies Store (Fig. 15), on Loramies Creek.

[3] "The surveys of the Military land are in good train Martin & Biggs have been in the woods more than a month, Jacksons set out the 13th instant and Mathews starts next Monday." (Putnam to Secretary of the Treasury, Marietta, July 22, 1797, ibid.)

[4] Ludlow, assigned one of five districts into which the U.S. Military Reserve was divided for surveying purposes, had completed his field work by early March, 1798. (Putnam to Secretary of the Treasury, Marietta, March 9, 1798, ibid.) By the beginning of the survey season a year later Ludlow was well along in the subdivision of lands west of the Great Miami. (Putnam to Secretary of the Treasury, Philadelphia, March 15, 1799, ibid.)

[5] Putnam to Secretary of the Treasury, Marietta, September 18, 1799, ibid. A walnut stake, unearthed at Ludlow's turning point at Fort Recovery, may be seen at the Fort Recovery Museum.

[6] Putnam to Secretary of the Treasury, Marietta, May 6, 1797, and Putnam to Secretary of the Treasury, Philadelphia, March 15, 1799, ibid.

[7] Ibid.

Within the limits of the Greenville Treaty Line, Putnam assigned his deputy surveyors, once the military tract had been surveyed,[1] to the subdivision of land in the three locations specified by the Land Act of 1796:

(1) North of the Geographer's Line (Fig. 15), where surveying was performed in 1799.[2] Building upon the Geographer's Line, surveyors extended their township lines as far west as the Tuscarawas River, and as far north as the forty-first parallel, which had by now been demarcated as a base line for surveying in the Western Reserve.[3]

(2) West of the Seven Ranges (Fig. 15), where surveying occurred in 1798 and 1799.[4] Practically surrounded by areas already subdivided, surveying here was founded in a poorly coordinated fashion by several surveyors of the military tract, respectively.[5] On the West, township surveying terminated abruptly at the Scioto River, where Virginia's preemption began.[6]

(3) West of the Symmes Purchase (Fig. 15), where Israel Ludlow, in the fall of 1798, entered upon the first independent organization of a field of federal surveying since Thomas Hutchins' initiation of the Geographer's Line, in 1785. Beginning at the mouth of the Great Miami River, Ludlow directed a line northward, for the guidance of his other lines of subdivision.[7] Building upon this central meridian, he had gone

---

[1] Within the U.S. Military Reserve were the three small tracts dedicated to the use of Moravian missionaries to the Delaware Indians (Fig. 15). Survey of these three tracts, surrounding the communities of Schoenbrun, Gnadenhutten and Salem, was the first assignment to be completed under Putnam. (Putnam to Secretary of the Treasury, Marietta, July 22, 1797, ibid.) All returns on the subdivision of the Military Reserve itself were in the hands of Putnam by the middle of November, 1798. (Putnam to Secretary of the Treasury, Marietta, November 13, 1798, ibid.)

[2] Putnam to Secretary of the Treasury, Marietta, October 24, 1799, ibid.

[3] For the field notes of this surveying, transcribed and combined with notes for subsequent subdivision, see Ohio Field Notes, Records of the General Land Office (Record Group 49), Cartographic Records Branch, the National Archives.

[4] Putnam to Secretary of the Treasury, Marietta, March 9, 1798, and October 24, 1799, Letters Received from the Surveyor General, Northwest Territory.

[5] The separated nature of surveying here is made evident by the non-accordant lines shown on "Map of Ohio Showing Original Land Subdivisions," accompanying Sherman, Ohio Land Subdivisions. For field notes, transcribed and combined with notes for subsequent subdivision, see Ohio Field Notes.

[6] Virginia surveyors had been locating claims between the Scioto and the Little Miami rivers since 1787, and Virginia's soldiers had been able to secure title from the United States to their claims in this area since 1790. (Hutchinson, "Bounty Lands of the American Revolution," pp. 51-56.)

[7] For record of Ludlow's survey of this line, already

far toward filling in the gap between the Great Miami River on the east and the Greeneville Treaty Line on the west one year after initiating the survey.[1]

As the preparation of returns on surveying in these three localities[2] neared completion, early in the year 1800, the end of Rufus Putnam's responsibility under the Land Act of 1796 was at hand. We shall now consider the status of the rectangular land survey system at this time.

---

scheduled to be a state boundary (Fig. 6), see transcribed notes, presented township by township, Ohio Field Notes.

[1] Putnam to Secretary of the Treasury, Marietta, October 24, 1799, Letters Received from the Surveyor General, Northwest Territory.

[2] For further discussion of subdivision in these three localities, see Sherman, Ohio Land Subdivisions, pp. 113, 115-127.

CHAPTER XI

STATUS OF THE AMERICAN RECTANGULAR LAND

SURVEY SYSTEM IN 1800

Having brought our story down to 1800, we have arrived at the chronological limit of this study: the end of the beginning of the U.S. public land surveys. Immediately in prospect, as of the opening of the year 1800, was the passage of a new act in Congress amending the Land Act of 1796.[1] Under its terms, land sales soon began,[2] and Putnam found himself contracting for the subdivision of all townships hitherto left whole. Some townships were to be divided into half-sections, in keeping with a provision brought forward in both 1785 and 1796 and now passed into law.[3] Later land laws required the subdivision of land into even smaller tracts, but the basic pattern of boundaries remained undisturbed.[4] By 1800, the American rectangular land survey system had become firmly established.

Passing over the general subject of land disposal, and

---

[1] For this amendatory act, approved May 10, 1800, see Carter, Territorial Papers, III, 88-97.

[2] Land offices were opened at Steubenville, in the Seven Ranges area, at Marietta, where the economic and political affairs of the Ohio Company settlements were centered, at Chillicothe, the "capital" of the Virginia Military District, and at Cincinnati, the principal city of the Symmes Purchase (Fig. 15). By the end of 1801, sales at these four offices totalled nearly 400,000 acres.

[3] The law directed that, of the alternate townships left unsubdivided (and scheduled for sale by quarter-section) under the Land Act of 1796, those east of the Muskingum River (Fig. 15) should be laid out in sections after the fashion specified in the Land Act of 1796 (Fig. 16A). Those west of the Muskingum were to be divided into half-sections by the running of lines at one-mile intervals and the placement of posts at half-mile intervals, as shown, Fig. 16C. Note that the Seven Ranges fell in the former, and the area west of the Symmes Purchase in the latter category.

[4] A land law of 1805 called for the subdivision of public lands into quarter-sections (United States Statutes at Large, II, 313), a land law of 1820 provided for the halving of quarter-sections (ibid., III, 566), and finally a law of 1832 authorized the laying out of quarter-quarter-sections by federal surveyors (ibid., IV, 503).

confining attention to surveying, as has been customary throughout this study, the present chapter will describe the organization of survey work in 1800, under the following heads: (1) chain of command, (2) the contract system, (3) base lines and principal meridians, (4) general surveying procedure, and (5) survey records. In general, a forward-looking view will be adopted, relating conditions of 1800 to those of later periods. Summary and evaluation of developments up to 1800 will be reserved for the ensuing, final chapter.

Chain of Command

The highest authority directly concerned with the progress of surveying, in 1800, was the Secretary of the Treasury, by virtue of his general responsibility for the sources of the government's income. The holder of this office, Oliver Wolcott, had drawn up the first set of general instructions covering federal surveying policy, in 1797.[1] If Wolcott took an active interest in the surveys, his successor, the highly capable Jeffersonian Albert Gallatin, was even more attentive to their programs. For more than a decade after the Spring of 1801 Gallatin kept in continuously close touch with operations in the field.[2] Late in his term of office, in 1812, part of the responsibility for the administration of public land policy passed to a newly created bureau within his department, the General Land Office.[3] The appointment of Edward Tiffin as the first Commissioner of the General Land Office[4] marked the beginning of the development of that bureau into a

---

[1] These instructions appear in Secretary of the Treasury to Rufus Putnam, in Carter, Territorial Papers, II, 591-594.

[2] Gallatin became Secretary of the Treasury in 1801, after a brief interim period of service by Samuel Dexter. Gallatin's active interest in public land surveying in the Northwest may be followed in correspondence in Carter, Territorial Papers, II, VII, X, and XVI, passim.

[3] The General Land Office, created by Act of Congress April 25, 1812, functioned at first as an agency for central control of land disposal, with little regard to surveying. (Milton Conover, The General Land Office: Its History, Activities and Organization [Baltimore: The Johns Hopkins Press, 1923], p. 16.)

[4] Tiffin, who had already served as Ohio's first governor, is credited with having collected and organized papers concerning the public lands, and with having saved these records from the risk of destruction by the British in 1814, in William E. Gilmore, Life of Edward Tiffin, First Governor of Ohio (Chillicothe, Ohio: Horney & Son, 1897), pp. 122-128.

famous American institution which for more than a century presided over the disposition of the public domain.[1]

Next below Wolcott, in the chain of command effective in 1800, was Rufus Putnam, who as Surveyor General was to serve only three more years, until removed by President Jefferson.[2] The office of Surveyor General, in Putnam's time, had not achieved the position of power which it was to enjoy during the early years of the General Land Office, nor was it an office with more than one incumbent, as it was shortly to become. Putnam's successor, Jared Mansfield, shared his title with the Surveyor General of lands south of Tennessee, and the number of surveyors general continued to increase from Mansfield's time onward, as the field of survey took on continental dimensions. From the outset, a problem in filling the office of Surveyor General was that of finding a man both capable and resistant to opportunities inherent in the office for unauthorized dealings in land.[3]

Directly under Putnam's control were deputy surveyors, of whom nine had been appointed under the Land Act, by 1800.[4] All of these men, like Wolcott and Putnam, were destined to pass from

---

[1] The authority of the General Land Office was extended over surveyors general in 1836. Transferred to the Department of the Interior in 1849, the General Land Office continued in operation until 1946, when its functions were taken over by the newly created Bureau of Land Management.

[2] For notice of Putnam's removal, see Gallatin to Putnam, Treasury Department, September 21, 1803, in Buell, Memoirs of Putnam, pp. 439-440. Mansfield's assumption of the duties of office at Marietta, November 1, 1803, is noted in Putnam to Gallatin, Marietta, February 18, 1804, ibid., p. 440.

[3] In 1796, the Secretary of State, noting the relatively low pay allowed a Surveyor General by Congress, wrote, "It may be doubted whether any very competent person for surveyor general can be found who will not improve the opportunity presented by his station of making or advancing his fortune in lands." (Timothy Pickering to President Washington, July 19, 1796, Miscellaneous Letters, Dept. of State, May-July, 1796.)

[4] These men were John Bever, Zaccheus Biggs, Ebenezer Buckingham, John Jackson, Elias Langham, Israel Ludlow, John Mathews, Levi Whipple and Thomas Worthington. (Note the reappearance not only of Ludlow but of Mathews, a veteran of surveying in the Ohio Company's lands who had been a chainman in the Seven Ranges.) In addition, two men, William R. Putnam, son of Rufus Putnam, and Absalom Martin, were among the deputies assigned to subdivision of the U.S. Military Reserve. Martin, having served under Hutchins in the Seven Ranges, was engaged by Putnam to survey the lands of French settlers of Gallipolis in 1796, and then, after Putnam became Surveyor General, to survey the military lands.

the surveying scene within a few years, but they appear to have
been representative of the type of frontiersman associated with
federal surveyorships for more than a century thereafter. Such
evidence as is available indicate that they were literate, generally unattached men, with prior experience in surveying, who
were ready to face the hardships of life in the field in exchange
for the income which employment by the federal government would
bring.[1]

In the employ of each deputy surveyor were chainmen and
other assistants. Their status, being dependent upon the operation of the contract system under which the deputy surveyors functioned, will be described in the discussion of that system, below.

The Contract System

Secretary Wolcott, having left to Putnam the determination of the terms under which surveyors were to be employed, in
his instructions of 1797, shortly received contracts which Putnam had concluded with the first two deputy surveyors engaged for
service.[2] Through these instruments and others which followed,
Putnam made his single lasting contribution to the traditions of
public land surveying. He was simply continuing, of course, a
procedure to which he had become accustomed as Superintendent of
Surveys for the Ohio Company.

---

[1] An interest in spying out the land also played its part.
Just as Winthrop Sargent sought further employment in public land
surveying in 1788, despite his position as Secretary of the Northwest Territory (Pershing, "Winthrop Sargent," pp. 49-50), so
Thomas Worthington, already a political figure of importance and
soon to be a United States Senator from Ohio, took up a surveyorship in 1799. Each man at the time in question was busy building
up a landed estate. By way of further advantage in land speculation, Biggs, Ludlow and Worthington became officials in local land
offices in 1800. (Secretary of the Treasury to Rufus Putnam,
Treasury Department, May 23, 1800, Letters Sent to Surveyors General, Records of the General Land Office [Record Group 49], Interior Section, Natural Resources Records Branch, the National Archives.)

[2] Wolcott expected contracts of some description (Carter,
Territorial Papers, II, 591, 592), and those which Putnam submitted may have been in general form little different from what
was anticipated, but the idea of contracting exclusively with
deputy surveyors, leaving the latter to undertake the hire of
their own survey crews, was obviously not in Wolcott's mind. The
first two contracts--with Absalom Martin and Zaccheus Biggs, for
subdivision of part of the Military Reserve--were enclosed in Putnam to Secretary of the Treasury, Marietta, May 18, 1797, to be
found in Letters Received from the Surveyor General, Northwest
Territory.

Putnam's contracts, which assigned a definite tract of country to each surveyor,[1] solved a financial problem which had plagued the government during the survey of the Seven Ranges, by absolutely limiting the claims of surveyors against the government to a certain compensation for each mile "run with the Compass and Measured with the Chain,"[2] and held the surveyors accountable for the expense of rectifying any errors discovered in their work.[3] Putnam assumed the same personal risk as that which Hutchins had once taken, by advancing credit to the surveyors on his own account.[4] The surveyors, in turn, hired their own parties of chainmen, axe-men, and whatever other helpers were needed. These helpers, and the surveyors themselves, were required to take an oath to perform their duties faithfully.[5]

Such was the nature of the contract system, under which deputy surveyors executed public land surveying from Putnam's time to the year 1910, when engineers directly employed by the federal government took over their duties.[6]

---

[1] Israel Putnam, for example, was assigned to the tract "bounded Southerly by the Ohio River Eastwardly by the Ohio and Great Miami River which runs into the Ohio--Northwardly and Westwardly by the Indian boundary." (Contract of May 31, 1798, accompanying Putnam to Secretary of the Treasury, Marietta, June 8, 1798, ibid.)

[2] Ludlow was allowed three dollars per mile, "as a full compensation for the whole expense in surveying, protracting and making the necessary returns and plans." (Ibid.) Ludlow, by the way, was permitted to "farm out" work to such other surveyors as Putnam might "approve and deputize for the purpose." (Ibid.)

[3] Ludlow, again, agreed that should any errors be committed or neglect take place in the execution of his work, Putnam would have the right "to cause the same to be rectified at the expence of the said Ludlow." (Ibid.)

[4] Wolcott had instructed Putnam that advances to the surveyors would have to be made on Putnam's own account. (Carter, Territorial Papers, II, 591.) Toward the end of the first survey season Putnam reported that he had already advanced "very considerable sums" so that the surveyors could get under way. (Putnam to Secretary of the Treasury, Marietta, October 7, 1797, Letters Received from the Surveyor General, Northwest Territory.

[5] This oath-taking was required by the Land Act of 1796. (Carter, Territorial Papers, II, 557.)

[6] An Interior Department appropriation act of June, 1910, terminated the contract system by providing that future public land surveying should be executed by "such competent surveyors as the Secretary of the Interior may select, at such compensation . . . as he may prescribe." (United States Statutes at Large, XXXVI, 741.)

Surveying: Base Lines and Principal Meridians

A need which remained unsatisfied in 1800 was that of providing a master framework for the surveyed townships. The need existed for two reasons. First, the conflict between rectangularity and convergency, which had been inherent in the Jefferson-Williamson plan of 1784, was still present in the Land Act of 1796. Second, the Land Act gave no guidance for convenient numbering of townships, as has been indicated. In an early chapter of this study, advance notice was given of a framework (Fig. 8) which eventually offered a practical solution to both of these difficulties.[1] Such a framework was established in 1804 by Jared Mansfield, Putnam's successor, with the laying down of an arbitrary meridian and an intersecting east-west base line, in present-day southern Indiana (Fig. 16D).[2] Mansfield progressively numbered his ranges of townships eastward and westward from the meridian, and the townships within each range northward and southward from the base line (Fig. 16D). The conflict between rectangularity and convergency was solved in a manner to be described below.

It should be explained again that the conflict between rectangularity and convergency consisted of this: north-south lines, which are not parallel by definition, were called for by law as the boundaries of all townships, yet at the same time the townships were required to be of constant width, so as to retain their rectangularity and uniformity of size. A compromise was necessary, and Mansfield's intersecting master lines provided a basis for it. By beginning at intervals of six miles, measured from the central--or "principal"--meridian, independent meridians could be laid down as township boundaries. To prevent these boundaries, when prolonged, from affecting the widths of the townships excessively, new, supplementary base lines could be run out from the principal meridian. Along these supplementary base lines, new north-south lines could be initiated, at the proper

---

[1] See p. 53, above.

[2] Mansfield's announcement of the laying down of the base line and of his plan for the placement of the principal meridian may be found in Mansfield to Secretary of the Treasury, Vincennes, October 26, 1804, in Carter, Territorial Papers, VII, 231-233. For interesting comments on the locating of this base line and principal meridian, based on original survey notes, see George R. Wilson, "Early Indiana Trails and Surveys," Indiana Historical Society Publications, VI (1919), 414-418.

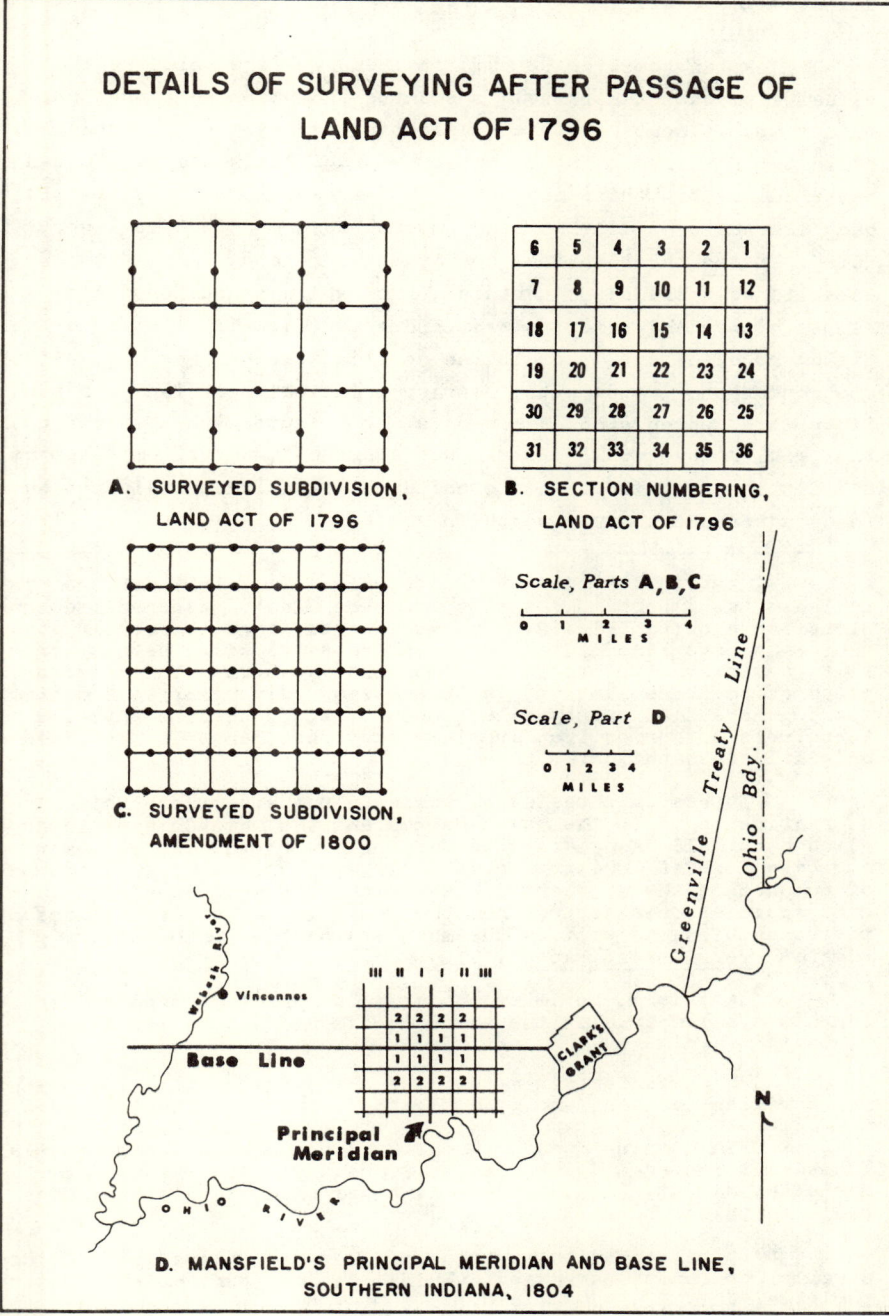

Fig. 16

six-mile intervals (Fig. 8). Thus was the necessary compromise made possible.[1]

Getting back to Putnam, in 1800, we find that the individual elements of Mansfield's eventual solution were present at this time, without anyone's envisioning the system into which they could be organized. In the military tract, for example, Putnam numbered his townships northward from a base line, only to fall back elsewhere on ill-matched numbers based on the Ohio River (Fig. 15).[2] In the tract west of the Symmes Purchase, Israel Ludlow established a principal meridian, from which he numbered his ranges of townships to the east and west (Fig. 15), but he established no base line.[3] As to the conflict between rectangularity and convergency, apparently neither principal meridians nor base lines were appreciated as potential contributors to a solution of the problem involved.[4] Putnam was aware of the problem,[5] but he had little reason to be concerned about it, since he did not enforce adherence to true north in surveying.[6]

---

[1] Mansfield's pioneer construction (Fig. 16D) differed in at least two important respects from the final, standardized construction shown in Fig. 8. First, its meridians, correctly spaced along the base line, diverged to an excessive width as they travelled southward (theoretically, at least), whereas the meridians south of the base line in Fig. 8 are shown diverging to a correct spacing more than twenty miles south of the base line. Second, there were no corrective, supplementary base lines in Mansfield's original construction.

[2] It was this basing of township numbers on the Ohio River--a practice begun in the Seven Ranges and the Ohio Company lands--which Mansfield was pleased to terminate with the establishment of his base line of 1804, remarking, "It would preserve an uniformity of numbers in the adjacent Townships of different ranges, & the mind would at once derive from the general plan a correct idea of their position, as well as the meanders of the [Ohio] River." (Carter, Territorial Papers, VII, 233.)

[3] Mansfield, in deference to this meridian, proposed calling his own comparable line the Second Meridian, as was done. (Ibid.)

[4] This was certainly the case where a base line and principal meridian had been employed together, in the Symmes Purchase.

[5] "In running North," wrote Putnam, "what we call parallel lines will converge." (Putnam to Secretary of the Treasury, Marietta, May 20, 1801, Letters Received from the Surveyor General, Northwest Territory.)

[6] Putnam aimed to organize his surveys generally by "lines parallel to former surveys." (Putnam to Secretary of the Treasury, Marietta, July 22, 1797, ibid.)

Special interest attaches to two of the meridians employed during Putnam's regime, since both coincided with state boundaries. The first of these was Pennsylvania's state boundary, originally intended for Hutchins' use but not brought into play as a reference line until Putnam extended surveying north of the Geographer's Line, in 1799.[1] The second was Ludlow's meridian, west of the Symmes Purchase, which coincided with a state line specified in the Northwest Ordinance of 1787 (Fig. 6). It is the eastern boundary of present-day Indiana.[2] The significance of these two lines lies in the fact that they briefly kept alive Jefferson's idea, embodied in his Ordinance of 1784, that state boundaries should serve as a controlling framework for public land surveying.

Secretary of the Treasury Gallatin intended to continue indefinitely the policy of using political boundaries for surveying purposes. In issuing instructions to the newly appointed Surveyor General of lands south of Tennessee, in 1803, Gallatin directed that the already-surveyed boundary with Spanish Florida (the southern boundary of parts of present-day Mississippi and Alabama) be used as a base line.[3] This was done. But Gallatin failed to impress Mansfield, Surveyor General Northwest of the Ohio, that he wanted the western boundary of present-day Indiana to serve as a principal meridian in the same manner as its eastern boundary had served.[4] Too late, he discovered that Mansfield had established a principal meridian in the midst of the prospective state. Mansfield had acted in innocence, not even knowing at this time that state boundaries had been specified in the North-

---

[1] Putnam directed that instruments be referred to the Pennsylvania boundary in the survey of this area, but results were not satisfactory. (Putnam to Secretary of the Treasury, Marietta, May 20, 1801, ibid.)

[2] Here alone, in the surveying undertaken prior to the advent of Mansfield, was a meridian employed as what Mansfield called "a Directrix from which the Ranges on each side of it may [be] counted." (Carter, Territorial Papers, VII, 232.)

[3] Secretary of the Treasury to Isaac Briggs, Treasury Department, April 8, 1803, ibid., V, 207-210. Gallatin also directed that a principal meridian be selected for the numbering of ranges of townships.

[4] For expression of Gallatin's disappointment see Secretary of the Treasury to Jared Mansfield, Treasury Department, March 13, 1805, ibid., VII, 268.

west Ordinance,[1] but as it transpired he was opposed to the idea of adjusting public land survey lines to state boundaries. He wanted townships to spread freely into all quarters of the compass from conveniently located initial points, without regard to state boundaries.[2] Converting Gallatin to his opinion, Mansfield set a precedent which was later followed elsewhere in the public domain. Thereafter, prime meridians and base lines coincided with state boundaries only where it was found convenient for them to do so (Frontispiece).

Surveying: General Procedure

Up to 1800, surveying consisted of the laying out of townships and the subdivision of them by lines run at intervals of two miles, obedient to the Land Act of 1796 (Fig. 16A). The general procedure followed in this work represented a continuation of practices instituted under the Land Ordinance of 1785, with a few improvements.

For measuring distances, the surveyors used Gunter's chains, as they and their successors would continue to do for a century thereafter.[3] An improvement in linear measurement introduced during Putnam's regime was the use of a standard chain, sent by Secretary of the Treasury Wolcott to Putnam in 1797.[4] Desirable as this standard chain was, as a basis for adjusting the lengths of the several chains employed in the field, reference to it at the outset of each deputy's surveying assignment could not insure against alteration in the length of a chain in the course of surveying, nor against errors due to carelessness. None the less, under Putnam's administration accuracy in chaining was somewhat greater than that which characterized surveying in the Seven Ranges.[5]

---

[1] Jared Mansfield to Secretary of the Treasury, Marietta, April 3, 1805, ibid., p. 277.

[2] Jared Mansfield to Secretary of the Treasury, Cincinnati, May 24, 1805, ibid., pp. 289-290.

[3] On the Gunter's chain, see pp. 74-75, above.

[4] Putnam to Secretary of the Treasury, Marietta, June 1, 1797, Letters Received from the Surveyor General, Northwest Territory.

[5] True, Putnam confided that he did not "think it necessary to oblige the Deputies to resurvey a line where they varied a few links or even chains in closing." (Putnam to Secretary of

For measuring directions, the surveyors continued to rely upon the circumferentor, an instrument with a magnetic compass needle which was to enjoy exclusive employment until the advent of the solar compass, in the 1830's.[1] The circumferentor was the instrument by which Congress expected the surveyors to run their boundaries "according to the true meridian," with the understanding that its magnetic readings would be used, in combination with an appropriate correction. Putnam, however, took liberties with the law. Taking into account the variation among individual compasses, he initiated surveying west of the Seven Ranges by having compasses "rectifi'd to the west boundary of the Seventh Range," but he required the application of no further correction, as would have been necessary for the establishment of true north-south lines.[2] In the area north of the Geographer's Line, Putnam required that compasses be adjusted to the western boundary of Pennsylvania--a true meridian--but no measures were taken to correct for either local attraction of the magnetic needle or the shift in the direction of magnetic north attendant upon the surveyors' removal westward from this initial line.[3] In the tract west of the Symmes Purchase, these same two sources of error were allowed to take effect.[4] Putnam, who made no secret of his non-adherence to true north, appealed to Congress without success for repeal of the true meridian clause in the Land Act of 1796.[5] Evidently unskilled in methods of determining true north, Putnam was in no position to enforce this part of the Land Act, and so it remained a practical nullity during his tenure of office. Enforcement waited upon the appointment of Jared Mansfield as Putnam's successor.

---

the Treasury, Marietta, August 10, 1801, ibid.) The judgment on improvement in chaining is based upon an examination of notes on closures in Ohio Field Notes, Records of the General Land Office (Record Group 49), Cartographic Records Branch, the National Archives.

[1] On the circumferentor, see pp. 77-78, above.

[2] Putnam to Secretary of the Treasury, Marietta, July 22, 1797, Letters Received from the Surveyor General, Northwest Territory.

[3] Putnam to Secretary of the Treasury, Marietta, May 20, 1801, ibid.

[4] Ibid.

[5] Putnam to Congress, Marietta, March 10, 1798, in American State Papers, Public Lands, I, 83.

If, in marking "line trees" and "corner trees" and in witnessing corner posts through the use of "bearing trees," the surveyors merely continued practices begun in the Seven Ranges, they showed improvement in the attention which they gave to the problem of closing surveyed lines upon one another.[1] This problem of closure, however, had not been met in any well-defined, systematic fashion as of 1800. A standard procedure began to take shape before Putnam left office, and continued to develop during Mansfield's regime.[2] Finally, in 1815, Edward Tiffin, having resigned as Commissioner of the General Land Office to become Surveyor General Northwest of the Ohio, issued a set of instructions which stipulated definite standards of performance--instructions which were the first in a series of directives extending down to the present day.[3]

Survey Records

Records of surveying, as of 1800, had not quite assumed those characteristics of form and distribution which were to become a familiar feature of federal land operations during the nineteenth century. Notes, taken by the surveyors in the field, had been submitted to the Surveyor General, as directed by the Land Act, but several decades were to pass before copies of these notes would be prepared for forwarding to the national capital.[4]

---

[1] The survey notes bear out Putnam in his claim that great care was taken to have all the "variations and closing distances" noted by the surveyors. (Putnam to Secretary of the Treasury, Marietta, May 20, 1801, ibid.)

[2] The most important advance evident in the survey notes for the three areas under consideration here was the introduction of random lines in the course of further subdivision of these areas, 1805-1806. These random, or more accurately, provisional, lines were employed as "feelers" in the direction of closing corners, roughly in a fashion which soon became standard in public land surveying. For an idea of technical improvements brought about, beginning in 1800, through legislation, see convenient summary of laws in U.S. Department of the Interior, Bureau of Land Management, Manual of Surveying Instructions, 1947, pp. 460-461.

[3] A typed copy of the Tiffin directive is included in a chronologically arranged collection of public land surveying instructions in the Office of the Chief Cadastral Engineer, Bureau of Land Management, Washington, D.C. Tiffin's instructions are reproduced in Sherman, Ohio Land Subdivisions, pp. 193-201.

[4] The notes in the National Archives referred to earlier as "Ohio Field Notes," covering the three areas of public land surveying, 1798-1800, were transcribed from the originals and sub-

Plats, at a scale of two inches to the mile (a scale passed on to Putnam by Secretary of the Treasury Wolcott, after the example of plats from the Seven Ranges)[1] were being drawn up, but in duplicate rather than triplicate, as would later become customary.[2] Finally "descriptions" were being prepared, whose form Putnam had developed on the basis of a general requirement in the Land Act.[3] These consisted of extracts from the surveyors' notes whereby the interested person could identify section corners, and gain an idea of what Putnam called "the quality of the land &c" along section lines.[4] These descriptions, together with the plats they were designed to accompany, became the model for subsequent land office records throughout the public domain.[5]

---

mitted to the General Land Office during the decade after 1835. Putnam viewed his own failure to have copies of the notes made and posted to the Secretary of the Treasury as a regrettable noncompliance with Wolcott's instructions of 1797 made necessary by lack of clerical help. (Putnam to Secretary of the Treasury, Marietta, August 10, 1801, Letters Received from the Surveyor General, Northwest Territory.)

[1] Secretary of the Treasury to Putnam, Treasury Department, March 14, 1797, in Carter, *Territorial Papers*, II, 593. All records of the survey of the Seven Ranges, as Putnam later pointed out when asked to prepare a general plat covering that area, were in the hands of the Treasury Department.

[2] Putnam later explained that as of May, 1800, he had nearly completed "two books of plats containing 176 Townships each," one to be retained and one to be sent to the Secretary of the Treasury. (Putnam to Secretary of the Treasury, Marietta, May 20, 1801, Letters Received from the Surveyor General, Northwest Territory.)

[3] The Surveyor General, in the words of the Land Act, was to "cause a description of the whole lands surveyed" to be made on the basis of the surveyors' notes. (Carter, *Territorial Papers*, II, 554.)

[4] The quotation is from the letter which first speaks in detail of these descriptions: Putnam to Secretary of the Treasury, Marietta, May 20, 1801, Letters Received from the Surveyor General, Northwest Territory. Putnam later gave a full account, saying that the descriptions were of "the corners . . . the witness or bearing trees, the kind of wood, inches diameter, course they bear and distance from the post, the quality of land and kind of timber on each subdividing line as reported by the Surveyor." (Putnam to Secretary of the Treasury, Marietta, July 11, 1803, *ibid.*)

[5] The plats and descriptions which were nearly complete by May, 1800, had to be reorganized and added to, in keeping with the amendatory land law of 1800. Combined with other plats and descriptions they went into the formation of two new books which were finished by July, 1803 (*ibid.*). Of these two volumes, the

From the survey records at his disposal, Putnam first compiled general plats, one for the use of each of the land offices where sales were opened, 1800-1801.[1] Then, presumably as an extension of a map showing lands south and east of the Greenville Treaty Line,[2] he prepared a map exhibiting the full extent of the newly created State of Ohio.[3] This map, based upon Mathew Carey's published map of the Seven Ranges and wherever possible upon survey records in Putnam's office, was still compelled to rely in part upon Thomas Hutchins' map of the West, which antedated the era of federal public land surveying.[4]

The striking success of Putnam's map[5] lay in its rectifi-

---

one sent to the Secretary of the Treasury may be found in Records of the General Land Office (Record Group 49), Cartographic Records Branch, the National Archives. It is titled on backstrip "Ohio.Vol. 1," and on title page, "Record Book No. 1, for the Secretary of the Treasury." Anyone particularly interested in this precedent-setting volume should note that the descriptions, verso, would all face the plats to which they refer if pages were rearranged in accordance with Putnam's index on the title page. It should also be noted that the volume does not include Ludlow's work west of the Symmes Purchase, the returns from which were long delayed.(Putnam to Secretary of the Treasury, Marietta, August 10, 1801, Letters Received from the Surveyor General, Northwest Territory.)

[1] Copies of these general plats, for the land office districts of Steubenville, Marietta, Chillicothe and Cincinnati, as defined in the amendatory land law of 1800, were submitted to the Secretary of the Treasury in May, 1801. (Putnam to Secretary of the Treasury, Marietta, May 20, 1801, Letters Received from the Surveyor General, Northwest Territory.) What appear to be the general plats in question, for all but the Steubenville district, are in Old Map File, Records of the General Land Office (Record Group 49), Cartographic Records Branch, the National Archives.

[2] Such a map, an undated and untitled manuscript (scale, 1 inch to about 5 miles), may be found in Old Map File. (Ibid.)

[3] "Map of the State of Ohio by Rufus E. Putnam, Surveyor General of the United States," in Thaddeus M. Harris, The Journal of a Tour into the Territory Northwest of the Alleghany Mountains; Made in the Spring of the Year 1803 (Boston, 1805), foll. p. 271. This book is dedicated to Putnam, from whom the author says he learned much. (Ibid., pp. iii-iv.)

[4] See note on the face of "Map of the State of Ohio," ibid., and, respecting the debt to Carey's map, Putnam to Secretary of the Treasury, Marietta, May 20, 1801, Letters Received from the Surveyor General, Northwest Territory.

[5] Reproductions of the map in Harris' book, reduced in size, may be found in Magazine of Western History, VI (April, 1881), foll. p. 248, and in Sherman, Ohio Land Subdivisions, p. 109.

cation and elaboration of stream lines on the basis of township-and-section survey notes--a success anticipated by Carey's earlier work. Improvement upon Putnam's picture of Ohio by John Mansfield[1] led to the appearance, in 1815, of a Hough and Bourne map in which watercourses were shown in new detail and the usefulness of the one-mile grid in locating such features as roads was made evident.[2] From this time onward, for about a century, the records of public land surveying enjoyed a position of primacy among the sources relied upon by map publishers, for all parts of the United States into which the public land survey grid was introduced.

---

[1] John F. Mansfield, Map of the State of Ohio Taken from the Returns in the Office of the Surveyor General (1806). Scale: 1 inch to about 10 miles. Among the details shown here and missing from Putnam's map are "Congressional" township boundaries.

[2] B. Hough and A. Bourne, Map of the State of Ohio from Actual Survey (Philadelphia: B. Hough & A. Bourne and J. Melish, 1815). Scale: 1 inch to about 5 miles.

CHAPTER XII

SUMMARY OF FINDINGS

Looking back over the foregoing chapters, one readily recognizes four general subjects to which the principal findings of this study pertain. They are, the rectilinear grid, the execution of surveying, the founders of the survey system, and the contributions of surveying to mapping and historical knowledge. In the present chapter, findings are summarized accordingly, in four sections. A concluding section is given over to a review of early opinions for and against rectangular subdivision of the public domain.

The Rectilinear Grid

The rectilinear grid of U.S. public land surveying, as we know it today, exhibits characteristics which accumulated during the early years of the survey system. The orientation of the grid originated in a proposed land law of 1784 which required that claims in the national domain be bounded by lines directed to the cardinal points of the compass.[1] The fact that today's grid lines enclose squares is also attributable to this proposed law, under the terms of which squareness would have promoted the operations of decimal arithmetic. (Reformation of land measurement, based upon decimal progression, was implicit in the law.)[2] Townships, as the major squares of the grid, made their first appearance in the Land Ordinance of 1785, and division of townships into square-mile sections gained partial acceptance in the same law.[3]

---

[1] This requirement was a feature of the Jefferson-Williamson plan for land subdivision. The plan is the subject of chap. ii of the present study, pp. 37-67, above.

[2] The basic unit in this reformed system of land measurement was the geographical mile, on which see pp. 46-50, above. On the relation of squareness to the system, see p. 57, above.

[3] Not townships but hundreds were to comprise the major squares of the grid, in the plan of 1784 (pp. 43-46, above). On

The influence of New England land policy upon the national rectilinear grid was of mixed and limited effect. The idea of townships came from this source, of course, the offering of whole townships for sale having been expected to induce emigrants from the East to confederate for the purpose of purchasing and settling together.[1] This expectation apparently played no part, however, in the perpetuation of townships in national land law in 1796.[2] The dimensions of the national townships--six miles square--may have followed New England precedent, but curiously, New England delegates in Congress were said to have adhered strongly to the idea of seven-mile townships in the debates of 1785.[3] The strict form of the national grid not only originated outside of New England tradition, but had to be protected against attempts made by men from that region to alter it. Delegates to Congress from Massachusetts moved first to permit the use of natural boundaries for townships, and then to relax the requirement that the lines of the rectilinear grid be regulated according to true north.[4] As late as 1800, conformity to true north had failed to take effect, due to failure of enforcement on the part of a Surveyor General who was accustomed to New England surveying methods.[5]

Application of the principle of prior survey to the rectilinear grid was definitely due to New England influence.[6] The fact that this principle, which called for subdivision of the land in advance of sale at auction, failed to take complete effect until after 1800, was due to a persistent desire to curtail the cost of preliminary surveying. On this account, the boundaries of individual sections within the Seven Ranges were left to the determination of purchasers,[7] and again in the areas initially surveyed

---

the introduction of townships in 1785 and the significance of square-mile sections relative to them, see pp. 92-95, above.

[1] For quotation to this effect, see pp. 92-93, above.

[2] This change is noted, p. 195, above.

[3] On the proposal of and rejection of seven-mile townships, see p. 96, n. 3, above.

[4] See pp. 89-92, above.

[5] This was Rufus Putnam (p. 215, above).

[6] On the principle of prior survey, see pp. 39-40 and 86-88, above.

[7] Sections, by law, were shown on paper only (pp. 94-95, above).

under the Land Act of 1796 buyers were obliged to complete the enclosure of their lots.[1]

The rectilinear grid was not provided with a proper foundation for well-controlled extension across the public domain until 1804, when a base line and a principal meridian were established in the southern part of present-day Indiana.[2] Ironically, the need for such master lines had been anticipated in the proposed law of 1784, by then long forgotten, which first brought forward rectangular surveying as a means for subdividing national lands. According to that law, state boundaries consisting of parallels of latitude and meridians of longitude were to be employed in the course of land subdivision essentially as base lines and principal meridians.[3] These proposed state boundaries were not put into effect, and although boundaries often resembling them were later established elsewhere, the Indiana precedent of keeping state lines separate from base lines and principal meridians was followed generally across the breadth of the public domain (Frontispiece).[4]

Finally, two common misunderstandings about the rectilinear grid call for correction. It should be understood, first, that the base lines and principal meridians controlling the grid are not to be thought of as having been laid out with reference to the earth's equator or to the meridian passing through Greenwich, England.[5] Second, it should be recognized that the conflict inherent in the survey grid between rectangularity and convergency, though not

---

[1] Three corners of each section were marked on the ground under this law (p. 196, above).

[2] On the establishment of these master lines by Jared Mansfield, see p. 210, above.

[3] These state boundaries, specified in the Ordinance of 1784 and invoked as survey control lines in the proposed land ordinance intended to accompany it, are discussed, pp. 15-36, above. Their equivalence to base lines and principal meridians is pointed out, p. 53, above.

[4] That a coincidence between these two kinds of lines was officially intended at the time the Indiana precedent was set is indicated, p. 213, above.

[5] That these should be arbitrary lines, located simply with a view to surveying convenience, was a point of basic importance to Jared Mansfield, in 1804 (p. 214, above). The longitude and latitude of all of these lines has been determined, since the time of their establishment, and may be found in U.S. Department of the Interior, Bureau of Land Management, Manual of Surveying Instructions, 1947, p. 168.

acted upon, was fully appreciated by men concerned with the national land system prior to 1800.[1]

Execution of Surveying

In the story of the execution of early public land surveying we find confirmation of the general proposition that passage from the Confederation period to the early years of government under the Constitution was made smooth by the continued functioning of key government personnel.[2] We have watched one man, Israel Ludlow, progress from service in the survey of the Seven Ranges, through participation in two extensive private surveys, to the assumption of a major role in surveying under the Land Act of 1796.[3] As we have noticed, two other men engaged as deputy surveyors after passage of the Land Act of 1796 also served under the pre-Constitution Land Ordinance of 1785;[4] and Rufus Putnam, first Surveyor General, functioned under both laws, although his experience did not extend back into the Seven Ranges.[5]

The general nature of surveying up to 1800 may be simply described. At first, field work was to be placed in the hands of a group of "gentlemen surveyors," each officially representing a state of the United States, but even before completion of the Seven Ranges work had been largely taken over by men of no public importance, and such men generally filled the surveyorships thereafter.[6] First and last, the surveyors were unlicensed.[7] It was only

---

[1] The conflict is described, p. 51, and belief in its early recognition is supported by citations, pp. 52 and 212, above.

[2] This proposition is put forward by Merrill Jensen in his The New Nation, p. 348.

[3] On Ludlow in the Seven Ranges, see pp. 134 and 140-143, above. Ludlow's part in the boundary survey for the Ohio Company is described, pp. 174 and 177, and in the Symmes surveys, pp. 181-184, above. On his surveying after passage of the Land Act of 1796, see pp. 201-204, above.

[4] These men were John Mathews and Absalom Martin. On Mathews, see p. 161, n. 2 and p. 207, n. 4, above.

[5] Putnam had administered surveying under the Land Ordinance of 1785 as Superintendent of Surveys for the Ohio Company (p. 179, above).

[6] The surveyor's role as contemplated in the Land Ordinance of 1785 is described, pp. 101-102, above. On the contrasting function of the surveyor in a system based on land warrants, see pp. 70-71 and 183, above. For further on surveyors, see pp. 123-128, 133-134, 161, 198, and 207-208, above.

[7] Omission of any licensing requirement from national land

required that they appear competent to the director of surveys. For measuring directions, the surveyors employed an instrument called a circumferentor, consisting of a magnetic compass equipped with sight vanes and a mount.[1] For measuring distances, they employed a surveyor's chain.[2] In laying down the national rectilinear grid by these means, through the forests of the Ohio country, the surveyors performed in a manner which the modern observer is more inclined to excuse than admire. To judge by their records of survey, the surveyors made no contribution to the solution of the problem of closing traverse lines.[3] They oriented their lines to true north neither in the areas governed by the Land Ordinance of 1785, where they were exempted from the necessity of doing so, nor, in general, in those areas governed by the Land Act of 1796.[4] On the other hand, these pioneers set a lasting precedent in their manner of marking trees, and in their record keeping practices.[5]

The progress of surveying, 1785-1800, cannot be judged properly unless allowance is made for two more or less constant inhibitory influences. The first of these was an overriding desire for economy, and the second was a constant threat of Indian interference. In the Seven Ranges, it was the latter influence which interrupted and greatly delayed the progress of surveying,[6] and the former influence which accounted for the suspension of government surveying in favor of surveying by private land

---

began with the proposed land ordinance of 1784, which broke away from Virginia tradition in this respect, as pointed out, p. 72, above.

[1] The circumferentor is introduced, pp. 77-78, above.

[2] The surveyor's chain is introduced, pp. 74-75, above.

[3] The running of lines in the Seven Ranges is discussed, pp. 146-149, above. On the quality of surveying after passage of the Land Act of 1796, see pp. 214-216, above.

[4] The true north requirement, first brought forward in 1784 (p. 77, above), was enacted into law in 1785 (p. 96, above), repealed a year later (pp. 92, 97, above), and restored as a requirement in the Land Act of 1796 (pp. 196-197, above), only to be evaded for the ensuing few years (p. 215, above).

[5] This point is made, pp. 158-159, above, and further elaborated upon, p. 217, above.

[6] See pp. 128-130, 136-137 and 141, above. On the Indians' early land cessions, see pp. 10-15, and on their general deployment in 1785, pp. 113-116, above.

companies.[1] They combined to reduce Thomas Hutchins to a state of despair.[2] Not until after the Battle of Fallen Timbers, in 1794, was the danger of Indian attack removed.[3] Impairment of effectiveness due to financial limitations continued up to and beyond 1800.[4]

Two discoveries deserving of separate mention, with reference to the execution of surveying, pertain to the Ohio Company of Associates, an organization widely known for its colony founded in the country west of the Seven Ranges in 1788. First, the organizers of this company were clearly the major immediate beneficiaries of surveying in the Seven Ranges, what with the gathering of intelligence by five of their representatives, and the extension of the Army's protective influence down the Ohio River, in consequence of that surveying.[5] Second, the contract system, under which federal deputy surveyors subdivided the greater part of the U.S. public domain between 1797 and 1910, originated in a resolution of the Ohio Company.[6] It was brought over into federal administrative procedure by Rufus Putnam.[7]

The Founders

If the present study has thrown no new light on the origin of the idea of rectangular land subdivision, it has made apparent how honors should be distributed among the men responsible for adopting the idea and developing it into the American rectangular land survey system. Primary credit belongs to Thomas Jefferson and Hugh Williamson, who, as members of the Continental Congress, brought forward rectangular subdivision for application to the na-

---

[1] The expense of surveying is reviewed, pp. 149-154, above. Congress' loss of faith in the original surveying program is remarked upon, p. 140, above.

[2] See pp. 175-176, above.

[3] Events leading up to the Treaty of Greenville, which ended the Indian danger in the Ohio country, are described, pp. 187-188, above.

[4] The latest example cited in this study of the adverse effect of limited means appears on p. 216, n. 4, above.

[5] The relation of the Ohio Company to surveying in the Seven Ranges is described, pp. 160-163, above.

[6] See p. 179, above.

[7] See p. 208, above. The financial problems of pre-contract system surveyors are considered, pp. 151-153, above.

tional domain, in 1784.[1] Jefferson was also the author of the plan for state boundaries of 1784 which anticipated the use of principal meridians and base lines; and as President, in 1804, he appointed Jared Mansfield, who established the first set of such master lines. Williamson, too, appears to deserve added recognition. Perpetuation of the rectangular survey system in the Land Act of 1796 seems not to have been based directly upon the Land Ordinance of 1785, but upon derivatives from it, one of which was a land act proposed to Congress in 1792 by Williamson.[2]

Other legislators who should be counted among the founders of the survey system are Rufus King and William Grayson, who took leadership in the framing of the Land Ordinance of 1785, and Jonathan Havens and Albert Gallatin, who figured importantly in the passage of the Land Act of 1796. King was spokesman for New England, in 1785, and as such stood for adoption of the principle of prior survey.[3] Havens, in 1796, was spokesman for New York. By urging adoption of the rectangular survey system which prevailed in western New York, he was in effect advocating perpetuation of the design for surveying in the Land Ordinance of 1785.[4] Havens thereby provided the second of the two apparent connecting links with that earlier law, the first of which was Williamson's proposed bill of 1792. The role of both Grayson and Gallatin was that of expediter. In 1785, successful adaptation of the Jefferson-Williamson plan to the wishes of Congress was mainly attributable to Grayson's efforts.[5] In 1796, Gallatin came forward, as Grayson had done, in behalf of compromise for the purpose of passage of

---

[1] Williamson's right to share the honor of authorship with Jefferson is asserted, p. 38, above.

[2] The introduction of Williamson's bill of 1792 is noted, p. 186, and the apparent influence of the bill in 1796 is brought out, pp. 193 and 194, above.

[3] On King's part in the composition of the Land Ordinance of 1785, see pp. 85, 87 and 95, above.

[4] Havens spoke out both for the rectilinear grid and for the six-mile township, as stated above, pp. 193 and 194. The connection between the design of land subdivision in western New York State and the Land Ordinance of 1785 is pointed out, pp. 61 and 193, above.

[5] Grayson's role, identified on p. 85, above, is repeatedly made evident in the chapter on the Land Ordinance of 1785, pp. 82-104, above.

public land legislation.[1] Later, having succeeded Oliver Wolcott in the office of Secretary of the Treasury, Gallatin acted as chief executive responsible for the advancement of public land surveying.[2]

Three directors of surveying, each of whom has been mistakenly credited at one time or another with having originated the American rectangular survey system,[3] contributed to the development of the system during its early years. They are Thomas Hutchins, Geographer in charge of surveying under the Land Ordinance of 1785, Rufus Putnam, first Surveyor General under the Land Act of 1796, and his successor, Jared Mansfield. Taking the last-named first, Mansfield was not only the Surveyor General responsible for placing rectangular surveying upon a foundation of principal meridians and base lines, he was also the first director of surveying to envision the extension of rectangular surveying over a great area.[4] Putnam not only introduced the contract system into public land policy, but imparted to rectangular surveying during the first few years after its readoption in the Land Act of 1796, the prestige of his name, that of a pioneer leader in the settlement of the Northwest Territory.[5] Finally, Hutchins should be credited with responsibility for the field and record-keeping practices which passed into the tradition of public land surveying from the survey of the Seven Ranges. His position of honor as a founder is assured, by virtue of his having been first in the field in the cause of public land subdivision, but it should be realized that, in Hutchins' own view, his service in this connection simply delayed return to his true vocation in life--that of single-handed, large-scale exploration and mapping of the West.[6]

---

[1] See record of debates in 1796, as cited, p. 195, n. 5, above.

[2] See p. 206, above.

[3] These mistaken attributions are reviewed and corrected, pp. 54-66, above.

[4] That Mansfield expected a great expansion of public land surveying, and shaped his proposals for the design of the surveys accordingly, is demonstrated by a letter, cited, p. 214, n. 2, above.

[5] Putnam's eminence in the Ohio country is given recognition, pp. 179 and 200, above.

[6] Late in 1788, shortly before the end of his life, Hutchins thought he had found an opportunity for returning to this vocation (p. 175, above). Regarding the map of the West to which Hutchins

Contributions to Mapping and Historical Knowledge

The purpose of the surveying described in this study was the conversion of outlying wilderness into saleable parcels of property. Toward this end, surveyors not only subdivided the land but also recorded evidence of their work in the form of notes.[1] The content of the notes was made available to prospective purchasers, together with plats showing the surveyed boundaries.[2] Beyond their immediate value in the interest of sales, these records when new offered material for the use of map makers. Today, as any researcher may confirm for himself, they offer data for the increase of historical knowledge.

Exploitation of public land survey records for purposes of mapping began even before completion of the survey of the Seven Ranges. Using indications in the survey records of the position of streams at the points of their intersection with township boundaries, Manasseh Cutler was able to produce a rectified picture of watercourses in that part of his map of the Ohio country, of 1788, which covered the first four ranges of public land townships.[3] In 1796, by which time the survey grid had enabled the Ohio Company and the Symmes group of associates to produce maps of their respective territories, Mathew Carey published a map of the Seven Ranges showing in admirable detail a drainage pattern developed from survey records.[4] Using this map and survey records for additional areas wherever possible, Rufus Putnam prepared a map exhibiting the full extent of the newly created State of Ohio.[5] Putnam's map, in turn, started a tradition of basing the mapping of states upon public land survey records, a tradition which continued throughout the nineteenth century.[6]

---

owed his reputation prior to his public land surveying activity, see pp. 17-19, 26 and 123, above.

[1] On notes taken for the Seven Ranges, see pp. 138, 142-143, 159 and 165, above. On notes taken after requirement thereof in the Land Act of 1796 (p. 196, above), see p. 203, nn. 3, 5 and 7, and p. 216, above.

[2] During Putnam's regime the content of the notes began to be presented in the form of "descriptions" (p. 217, above). On plats, see pp. 75-77, 97, 139, 142-143, 159, 165 and 217, above.

[3] See pp. 164-165, above.   [4] See p. 165, above.

[5] See p. 218, above.

[6] The first maps to follow Putnam's map are cited, p. 219, above.

Of the historical evidence found in the survey records examined for the present study, that most likely to be of interest to researchers pertains to trees, trails, sites of human habitation, place names and land evaluation.[1] Information on the species, diameter and position of trees, set down for boundary identification purposes,[2] comprises the most satisfactory body of historical evidence left behind by the early surveyors. Due to this information, and to data of exactly the same kind produced by later surveyors, public land survey records have come to be regarded by twentieth century botanists, geographers and historians as a standard source of reference for use in forest reconstruction.[3]

Early Opinions for and against the Rectangular Survey System

In conclusion, it should be made clear that conflict of opinion over the desirability of the American rectangular land survey system is as old as the system itself. Opposition to rectangular subdivision of the public domain was first voiced by advocates of the Southern system of indiscriminate locations, among them George Washington, who declared, "To the end of time [the public lands] will not, by those who are acquainted therewith, be purchased either in Townships or by square miles."[4] However mistaken this expectation may have been,[5] an associated proposition, that lands ought to be subdivided with due regard for natural features, especially river bottoms, was never success-

---

[1] Under these heads, evidence in the survey records for the Seven Ranges is discussed, pp. 166-168, above. Much the same conclusions as those appearing there would have been set down, had discussion been extended to cover records down to 1800.

[2] On the use of trees in the marking of boundaries, see pp. 73-74, 97, 158 and 196, above.

[3] General acceptance of the records on this account became apparent to the present author in the course of preparing the following article: William D. Pattison, "Use of the U.S. Public Land Survey Plats and Notes as Descriptive Sources," *Professional Geographer*, New Series, VIII (January, 1956), 10-14.

[4] Washington to Grayson, Mount Vernon, August 22, 1785, in Fitzpatrick, *Writings of George Washington*, XXVIII, 234. The quotation appears earlier, p. 89, above.

[5] Whether because of rectangular subdivision or not, it is true that the first attempt at the sale of land by townships and sections was a failure. See pp. 155-157, above.

fully denied by members of Congress who favored rectangular surveying. This preference for the adjustment of property lines to the natural landscape, expressed in the debates of both 1785 and 1796, was shared by such leaders outside Congress as Arthur St. Clair,[1] first governor of the Northwest Territory, and the Geographer Thomas Hutchins.[2]

Adoption of rectangular surveying, in the face of opposition, was apparently due to the successful urging of two beliefs about the system: that it would be relatively inexpensive to put into effect, and that it would provide security of title for purchasers of the land. William Grayson, the man principally responsible for passage of the Land Ordinance of 1785, assured a correspondent that rectangular surveying "would be attended by the least possible expence," and this assurance was repeated in the debates of 1796.[3] The advantage of security of land title was pointed out by Hugh Williamson, when the Jefferson-Williamson plan was laid before Congress, in 1784. His declared expectation that an unvarying grid would restrain surveyors from fraudulent measurement and an overlapping of lots was echoed on later occasions.[4] Grayson added the belief that rectilinear boundaries once laid down were likely to endure,[5] and the committee which brought out the Land Act of 1796 worked on the assumption that Congress, by providing boundaries both strict and durable, could expect to raise the price at auction of public lands.[6]

---

[1] St. Clair favored transfer of the Pennsylvania system of indiscriminate locations (given incidental attention, p. 68, n. 2, above) to the national domain, in the place of rectangular surveys (St. Clair to John Jay, Fort Harmar, December 13, 1788, in Smith, St. Clair Papers, II, 104.)

[2] Hutchins favored the survey of rivers, whose courses would have served to bound or locate large properties. Referring to a plan which he submitted for such a survey, in 1788, Hutchins wrote, "Congress will be as effectually enabled to dispose of the Lands in the proposed plan with as much if not more satisfaction to themselves and purchasers than they now do, or will ever be able to do, under the present Ordinance." (Hutchins to Committee of Congress, New York, March 5, 1788, Papers Cont. Cong., LXXVIII, Pt. XII, 541.

[3] Grayson's assurance may be found in Grayson to Washington, New York, April 15, 1785, in Burnett, Letters of Members, VIII, 95. The quotation appears earlier, p. 88, above.

[4] Williamson's statement is quoted, pp. 40-41, above.

[5] See p. 88, above.   [6] See p. 193, above.

Neither the friends nor the foes of the rectangular survey system, in its early years, seem to have anticipated the great extent to which gridded property lines would eventually determine land use patterns (a dramatic effect noted in the opening sentence of the introduction to the present study),[1] but early forecasts were correct with respect to the ability of rectangular surveying to prevent controversy over boundaries. The modern case in favor of the system rests principally upon the fact that disputes over land title in the former public domain have been relatively few, in marked contrast to the amount of litigation which has arisen in those parts of the United States not favored with a basic rectangular land survey.[2]

---

[1] The effect of the original surveyed squares upon land use has not only been pointed to as an impressive fact, by such authors as those cited, p. 1, n. 1, above. By some writers it has been greatly deplored. Most recently, Hildegard Binder Johnson has brought serious charges against the rectangular survey system in her "Rational and Ecological Aspects of the Quarter Section: An Example from Minnesota," Geographical Review, XLVII (July, 1957), 346. Unecological treatment of both woods and soil by farmers is here declared to be a result of rectangular subdivision of the land.

[2] See, for example, George F. Tyrell, "Background and Development of Cadastral Surveys," Surveying and Mapping, XVII (January, February, March, 1957), 33.

APPENDIX

NOTES ON THE FIGURES

Frontispiece. Extent of the American Rectangular Land Survey System.
Base map: Goode's Series of Base Maps, United States, Number 110. Chicago: University of Chicago Press, 1937.
Data principally from U.S. Department of the Interior, General Land Office, United States, Showing Principal Meridians, Base Lines and Areas Governed Thereby. Washington: Government Printing Office, 1937. Scale: 1 inch to 125 miles.

Figure 1. Claims in the Northwest by Virginia, Connecticut and Massachusetts.
Base map: Charles O. Paullin, Atlas of the Historical Geography of the United States. Edited by John K. Wright. Washington: Carnegie Institution of Washington, 1932. Plate 47B.
Data principally from the plate cited as base map for this figure, and, in the same atlas, Plate 47C.

Figure 2. Lands Reserved in Virginia's Deed of Cession.
Base map: Same as map cited as source of data for Frontispiece.
Data principally from Virginia's Deed of Cession, as cited in the present study, p. 10, nn. 1-4.

Figure 3. Three Boundaries for Early Indian Cessions in the Northwest.
Base map: Erwin Raisz, Landforms of the United States. 6th ed. revised. Boston: Ginn and Co., 1954. Scale: 1 inch to about 75 miles.
Data principally from documents specifying the Indian cession boundaries concerned, as cited in the present study, p. 11, n. 4; p. 14, n. 4; and p. 15, n. 2.

Figure 4. Projected Boundaries for Western States, 1784.
Base map: Julian P. Boyd (ed.), The Papers of Thomas Jefferson. Princeton, New Jersey: Princeton University Press, 1950--. Vol. VI, Map III, p. 591.
Data from (1) map cited as base map for this figure, (2) Jefferson's plan for western states, as cited in the present study, p. 15, n. 5, and (3) Ordinance of 1784, as quoted in the present study, p. 16.

Figure 5. Boundaries for Western States, 1784, Shown on Contemporary Map Base.

Base map: Photograph, reduced in size, of Thomas Hutchins, A New Map of the Western Parts of Virginia, Pennsylvania, Maryland and North Carolina . . . . London: Engraved by I. Cheevers, 1778. Scale of original: 1 inch to about 20 miles.
Data from Ordinance of 1784, as quoted in the present study, p. 16.

Figure 6. Lines Prescribed by the Northwest Ordinance for Bounding States.

Base map: Goode's Series of Base Maps, United States, Number 310. Chicago: University of Chicago Press, 1938.
Data from the Northwest Ordinance, as cited in the present study, p. 35, nn. 2-4.

Figure 7. Jeffersonian Units of Land Subdivision.

Data from land ordinance proposed in 1784, as quoted in the present study, pp. 37-38.

Figure 8. Solution to the Problem of Rectangles and Meridians.

Base diagram and source of data: U.S. Department of the Interior, Bureau of Land Management, Manual of Instructions for the Study of the Public Lands of the United States, 1947. Washington: Government Printing Office, 1947. Fig. 16, p. 170.

Figure 9. Two Examples of Township-Bounding in New England.

Base map and source of data for Part A: U.S. Department of Commerce, Bureau of the Census, Maine: Minor Civil Divisions. Washington: Government Printing Office, 1934. Scale: 1 inch to 8 miles.
Data for Part B from U.S. Geological Survey, Topographic Map of the United States, Wilmington, Vermont Sheet. Washington: Government Printing Office, 1899. Scale: 1:62,500.

Figure 10. Numbering of Townships and Lots under the Land Ordinance of 1785.

Data for Part A from Land Ordinance of 1785, as cited in the present study, p. 98, nn. 1 and 2.
Data for Part B from surveyors' plats in National Archives, as cited in the present study, p. 98, n. 6.

Figure 11. The First Scene of Survey and Its Environs.

Base map: Same as base for Figure 3.
Data on broad physiographic areas from Nevin M. Fenneman, Physiography of the Eastern United States. New York: McGraw-Hill Book Company, 1938. Plates I and II.
Data on Pennsylvania Road principally from entries in Diary of Winthrop Sargent cited in the present study, p. 107, n. 3, and p. 108, nn. 1 and 5, and associated entries.
Data for inset from several sources, cited in the present study, p. 111, nn. 1-6; p. 112, nn. 3 and 5; p. 113, n. 1; and p. 117, nn. 1-4.

Figure 12. The Seven Ranges.

Base map and source of data: Photostat, reduced in size, of relevant part of "Map of Ohio Showing Original Land Subdivisions," accompanying Final Report, Ohio Cooperative Topographic Survey, Vol. III: C. E. Sherman, Original Ohio Land Subdivisions. Press of the Ohio State Reformatory, 1925. Scale of original: 1 inch to 6 miles.

Figure 13. Surveying Diagrams.

Data for Part A from U.S. Department of the Interior, Bureau of Land Management, Manual of Instructions for the Survey of the Public Lands of the United States, 1947. Washington: Government Printing Office, 1947. Fig. 15, p. 156.

Data for Part B from map compiled on the basis of sources cited in the present study, p. 148, n. 5.

Figure 14. Ohio Company Lands and the Miami Purchase.

Base map: "Ohio Country, 1787-1803," in James Truslow Adams (ed.), Atlas of American History. New York: Charles Scribner's Sons, 1943. Plate 85.

Data from plate cited as base map for this figure, and from Ohio Company documents cited in the present study, p. 174, n. 1, and p. 178, nn. 1-3.

Figure 15. The Three Areas First Surveyed under the Land Act of 1796.

Base map: Same as base map, Figure 14.

Data from map cited as base for this figure, and from map cited as source of data for Figure 12.

Figure 16. Details of Surveying after Passage of Land Act of 1796.

Data for Part A from Land Act of 1796, as cited in the present study, p. 196, n. 6.

Data for Part B from Land Act of 1796, as cited in the present study, p. 197, n. 5.

Data for Part C from Amendment to Land Act of 1796, as cited in the present study, p. 205, n. 3.

Base map and source of data for Part D: photostat, reduced in size, of John Collett, An Outline Map of Indiana. 1882. Scale of original: 1 inch to 25 miles.

BIBLIOGRAPHY

Manuscripts and Manuscript Collections

Chicago Historical Society, Chicago, Illinois.
> John Montgomery Papers.

Historical Society of Pennsylvania, Philadelphia, Pennsylvania.
> Thomas Hutchins Papers.
> Miscellaneous Collection.

Illinois Historical Survey, Urbana, Illinois.
> George Morgan Papers, letter, Thomas Hutchins to Committee of Congress, August, 1788. (Photostatic copy.)

Library of Congress, Manuscripts Division, Washington, D.C.
> Facsimiles from Spanish Archives, Archivo Historico Nacional.
> George Morgan Papers.
> Northwest Territory Papers, Miscellaneous, 1787-1789.
> Papers of George Washington, Applications for Office under President Washington.

Marietta College Library, Marietta, Ohio.
> Journal of John Mathews, July 10, 1786-April 21, 1787.

Massachusetts Historical Society, Boston, Massachusetts.
> Sargent Papers, Diary of Winthrop Sargent, June 18-December 21, 1786. (Typewritten copy.)

National Archives, Washington, D.C.
> Cartographic Records Branch: Records of the General Land Office (Record Group 49).
>> Field Notes of Symmes Purchase.
>> Ohio Field Notes.
>> Old Map File.
>> Plat Books: Ohio Vol. I and Ohio Vol. III.
>> Plats and Notes for the Seven Ranges.
>
> Natural Resources Records Branch, Interior Section: Records of the General Land Office (Record Group 49).
>> Letters Received from the Surveyor General, Northwest Territory.
>> Letters Sent to Surveyors General.
>> Records of Thomas Hutchins.
>
> Legislative, Judicial and Diplomatic Records Branch, Fiscal Section: Records of the General Accounting Office (Record Group 217).
>> Journal "C."
>> Journal "D."

Legislative, Judicial and Diplomatic Records Branch, Fiscal Section: Records of the Bureau of the Public Debt (Record Group 53).

    Journal, August 1, 1785-June 8, 1787.
    Journal, June 8, 1787-July 14, 1789.

Legislative, Judicial and Diplomatic Records Branch, Foreign Affairs Section: General Records of the Department of State (Record Group 59).

    Miscellaneous Letters, Department of State, 1796.

Legislative, Judicial and Diplomatic Records Branch, Foreign Affairs Section: General Records of the United States Government (Record Group 11).

    Papers of the Continental Congress.

Western Reserve Historical Society, Cleveland, Ohio.

    Samuel Parsons Papers.

    Jonathan Heart Papers.

    Miscellaneous Collection.

William L. Clements Library, Ann Arbor, Michigan.

    Papers of General Josiah Harmar, selected correspondence, May-July, 1787. (Photostatic copies.)

Published Collections

Acts and Laws of the Commonwealth of Massachusetts. 13 vols. Boston: Printed by Wright and Potter Printing Co., 1890-1898. Vol. I.

American State Papers: Documents, Legislative and Executive, of the Congress of the United States. 38 vols. Washington: Gales and Seaton, 1832-1861. Foreign Affairs, Vol. I; Indian Affairs, Vol. I; Public Lands, Vol. I.

Annals of Congress.

Batchellor, Albert S., et al. (eds.) New Hampshire Provincial and State Papers. 40 vols. Concord: State of New Hampshire, 1867-1943. Vols. XXIV, XXVII.

Bates, Samuel A. (ed.) Records of the Town of Braintree, 1640-1793. Randolph, Massachusetts, 1886.

Blume, F., Lachmann, K., and Rudorff, A. Die Schriften der romischen Feldmesser. 2 vols. Berlin: George Reiner, 1848-1852.

Bond, Beverley W., Jr. The Correspondence of John Cleves Symmes, Founder of the Miami Purchase. New York: The Macmillan Company, 1926.

Boyd, Julian P. (ed.) The Papers of Thomas Jefferson. Princeton, New Jersey: Princeton University Press, 1950--. Vols. II, IV, V, VI, VII, X.

Boyd, William K. (ed.) Some Eighteenth Century Tracts Concerning North Carolina. Raleigh, North Carolina: Edwards and Broughton Co., 1927.

Buell, Rowena (ed.). The Memoirs of Rufus Putnam and Certain Official Papers and Correspondence. Boston: Houghton, Mifflin and Company, 1903.

Burnett, Edmund C. (ed.) Letters of Members of the Continental Congress. 8 vols. Washington: Carnegie Institution of Washington, 1921-1936. Vols. VII, VIII.

Carter, Clarence E. (ed.) The Territorial Papers of the United States. Washington: Government Printing Office, 1934--.
    The volumes used in this study were:
        II-III. Territory Northwest of the River Ohio, 1787-1803.
          V. Territory of Mississippi, 1798-1817.
        VII. Territory of Indiana, 1800-1810.

Clark, Walter (ed.). The State Records of North Carolina. 16 vols. Goldsboro, North Carolina: Published under supervision of the Trustees of the Public Libraries, 1895-1907. Vols. XXIII, XXIV, XXVI.

Conway, Moncure D. The Writings of Thomas Paine. 4 vols. New York: G. P. Putnam's Sons, 1894-1896. Vol. II.

Cutler, William P., and Julia P. Life, Journals and Correspondence of Rev. Manasseh Cutler, LL.D. 2 vols. Cincinnati: Robert Clarke & Co., 1888.

Fitzpatrick, John C. (ed.) The Writings of George Washington from the Original Manuscript Sources, 1745-1799. 39 vols. Washington: Government Printing Office, 1931-1944. Vols. XXVII, XXVIII.

Ford, Worthington, C., et al. (eds.) Journals of the Continental Congress, 1774-1789, Edited from the Original Records in the Library of Congress. 34 vols. Washington: Government Printing Office, 1904-1937. Vols. XV, XVI, XVIII, XX, XXI, XXIII, XXIV, XXV, XXVI, XXVII, XXVIII, XXIX, XXX, XXXI, XXXII, XXXIII, XXXIV.

Hall, Charles S. Life and Letters of Samuel Holden Parsons. Binghamton, New York: Otseningo Publishing Co., 1905.

"Haldimand Papers," in Michigan Pioneer and Historical Collections, Vol. XX (1892).

Hamilton, Stanislaus M. (ed.) The Writings of James Monroe, Including a Collection of His Public and Private Papers and Correspondence Now for the First Time Printed. 7 vols. New York: G. P. Putnam's Sons, 1898-1903. Vol. I.

Hoadley, C. J. (ed.) Public Records of the Colony of Connecticut. 15 vols. Hartford: The Press of the Case, Lockwood and Brainard Co., 1850-1890. Vol. VIII.

Hulbert, Archer B. The Records of the Ohio Company. 2 vols. ("Marietta College Historical Collections, Ohio Company Series," Vols. I and II.) Marietta, Ohio: Marietta Historical Commission, 1917.

_____. Ohio in the Time of the Confederation. ("Marietta College Historical Collections, Ohio Company Series," Vol. III.) Marietta, Ohio: Marietta Historical Commission, 1918.

James, James A. (ed.) George Rogers Clark Papers, 1771-1781. (Collections of the Illinois Historical Library, Vol. VIII.) Springfield, Illinois: Illinois State Historical Library, 1912.

Journals of the American Congress, from 1774 to 1788. 4 vols. Washington: Way and Gideon, 1823. Vol. IV.

King, Charles R. (ed.) The Life and Correspondence of Rufus King, Comprising His Letters, Private and Official, His Public Documents and Speeches. 6 vols. New York: G. P. Putnam's Sons, 1894-1900. Vol. I.

Laws of the State of New York . . . from the First to the Twentieth Session, Inclusive. 3 vols. New York: Printed by Thomas Greenleaf, 1798. Vol. I.

Lipscomb, Andrew A., and Bergh, Albert E. (eds.) The Writings of Thomas Jefferson. 20 vols. Washington: The Thomas Jefferson Memorial Association, 1903. Vols. XII, XIII, XIV.

Mathews, Catherine V. C. Andrew Ellicott, His Life and Letters. New York: The Grafton Press, 1908.

Miller, Hunter (ed.). Treaties and Other International Acts of the United States of America. Washington: Government Printing Office, 1931--. Vol. II.

O'Callaghan, E. B. (ed.) Documents Relative to the Colonial History of the State of New York. Albany, New York: Weed, Parsons and Co., Printers, 1856-1861. Vol. VIII.

Original Land Titles in Delaware Commonly Known as the Duke of York Record. Wilmington, 1903.

Pennsylvania Archives, First Series. 12 vols. Philadelphia, 1852-1856. Vol. IX.

Pennsylvania Archives, Third Series. 30 vols. Harrisburg, 1894-1899. Vol. III.

Pickering, Octavius, and Upham, Charles W. The Life of Timothy Pickering. 4 vols. Boston: Little, Brown, and Company, 1867-1873. Vol. I.

Richardson, H. W., et al. (eds.) York Deeds. 18 vols. Bethel, Maine, 1903-1910. Vol. XVII.

Smith, William Henry. The Life and Public Services of Arthur St. Clair with His Correspondence and Other Papers. 2 vols. Cincinnati: Robert Clarke and Co., 1882.

Staples, William R. Rhode Island in the Continental Congress. Providence, Rhode Island, 1870.

Stetson, Charles W. Four Mile Run Land Grants. Washington: Mimeoform Press, 1935.

Thorpe, Francis N. The Federal and State Constitutions, Colonial Charters, and Other Organic Laws of the States, Territories and Colonies Now or Heretofore Forming the United States of America. (U.S. House Doc. No. 357, 59th Cong., 2d Sess.) 7 vols. Washington: Government Printing Office, 1909. Vols. III, VII.

Trask, William B., et al. (eds.) Suffolk Deeds. 14 vols. Boston, 1880-1906. Vol. X.

United States Statutes at Large.

Published Journals, Reports and Contemporary Writings

"Arthur Lee's Journal," Olden Time, II (July, August, 1847), 334-344.

Bond, Beverley W., Jr. (ed.) The Courses of the Ohio River Taken by Lt. T. Hutchins Anno 1766 and Two Accompanying Maps. Cincinnati: Historical and Philosophical Society of Ohio, 1942.

Bushnell, David I. (ed.) "Journal of Samuel Montgomery," Mississippi Valley Historical Review, II (September, 1915), 261-273.

Butterfield, Consul W. (ed.) Journal of Captain Jonathan Heart . . . to Which Is Added the Dickinson-Harmar Correspondence of 1784-5. Albany, New York: Joel Munsell's Sons, 1885.

Ellicott, Andrew. The Journal of Andrew Ellicott, Late Commissioner . . . for Determining the Boundary between the United States and the Possessions of His Catholic Majesty in America. Philadelphia, 1814.

Fitzpatrick, John C. (ed.) The Diaries of George Washington, 1748-1799. 4 vols. Boston: Houghton Mifflin Company, 1925.

Hutchins, Thomas. A Topographical Description of Virginia, Pennsylvania, Maryland, and North Carolina, Comprehending the Rivers Ohio, Kenhawa, Sioto, Cherokee, Wabash, Illinois, Mississippi, &c. . . . London, 1778.

———. A Topographical Description of Virginia, Pennsylvania, Maryland and North Carolina. Edited by Frederick C. Hicks. Cleveland: The Burrows Brothers Company, 1904.

Jefferson, Thomas. Notes on the State of Virginia. Edited by William Peden. Chapel Hill: University of North Carolina Press, 1955.

———. Report of the Secretary of State, on the Subject of Establishing a Uniformity in the Weights, Measures, and Coins of the United States. New York, 1790.

"Journal of General Butler," Olden Time, II (October, November, December, 1847), 433-464, 481-525, 529-531.

Mansfield, Jared. Essays Mathematical and Physical. . . . New Haven, Connecticut [1802].

Report of the Commissioners Appointed to Complete the Examination and Determination of All Questions of Title to Land . . . on the Isle of Martha's Vineyard. Boston, 1871.

Report of the Committee for the Sale of Eastern Lands: Containing Their Accounts from the 28th of October, 1783, to the 16th of June, 1795. [Boston, 1795.]

Report of the Secretary of Internal Affairs of the Commonwealth of Pennsylvania, Containing Reports of the Surveys and Re-Surveys of the Boundary Lines of the Commonwealth, Accompanied with Maps of the Same. Harrisburg, 1887.

Resolves of the General Court of the Commonwealth of Massachusetts Respecting the Sale of Eastern Lands; with the Reports of the Committees Appointed To Sell Said Lands; from March 1, 1781 to [June 22, 1803]. Boston, 1803.

Sargent, Winthrop. "List of Forest and Other Trees Northwest of the River Ohio," Memoirs of the American Academy of Arts and Sciences, II (1793), 156-159.

U. S. Congress, Senate. Report of the Commissioner of the General Land Office. (U. S. Senate Doc. No. 11, 25th Cong., 2d Sess.) Washington, 1837.

Washington, George. Journal of My Journey over the Mountains, While Surveying for Lord Fairfax, Baron of Cameron, in the Northern Neck of Virginia, Beyond the Blue Ridge, in 1747-8. Edited by J. M. Toner. Albany, New York: Joel Munsell's Sons, 1892.

Webster, Pelatiah. Political Essays on the Nature and Operation of Money, Public Finances, and Other Subjects. Philadelphia, 1791.

Williamson, Hugh. The History of North Carolina. 2 vols. Philadelphia: Thomas Dobson, 1812.

Surveying Instructions

Commonwealth of Massachusetts Land Court. Manual of Instructions for the Survey of Lands and Preparing Plans for the Land Court. Boston, 1913.

Davies, Charles. Elements of Surveying and Navigation. Rev. ed. New York: A. S. Barnes & Co., 1853.

Dodds, John S., et al. Original Instructions Governing Public Land Surveys of Iowa: A Guide to Their Use in Resurveys of Public Lands. Ames, Iowa: Iowa Engineering Society, 1943.

Flint, Abel. A System of Geometry and Trigonometry: Together with a Treatise on Surveying. 2d ed. Hartford, 1808.

Gibson, Robert. The Theory and Practice of Surveying. New York, 1821.

———. A Treatise of Practical Surveying; Which Is Demonstrated from Its First Principles. 5th ed. Philadelphia, 1789.

Love, John. The Whole Art of Surveying and Measuring of Land Made Easie. 3d ed. London, 1716.

Tracy, John Clayton. Surveying, Theory and Practice. New York: John Wiley & Sons, 1947.

U. S. Department of the Interior, Bureau of Land Management. Manual of Instructions for the Survey of the Public Lands of the United States, 1947. Washington: Government Printing Office, 1947.

U. S. Department of the Interior, General Land Office. Manual of Instructions for the Survey of the Public Lands of the United States, 1930. Washington: Government Printing Office, 1931.

———. Manual of Surveying Instructions for the Survey of the Public Lands of the United States and Private Land Claims. Washington: Government Printing Office, 1902.

Reference Works

Adams, J. Truslow (ed.). Dictionary of American History. 5 vols. New York: Charles Scribner's Sons, 1942.

Biographical Directory of the American Congress, 1774-1949. Washington: Government Printing Office, 1950.

Encyclopaedia Britannica. 11th and 14th eds.

Heitman, Francis B. *Historical Register of Officers of the Continental Army during the War of the Revolution.* Washington: The Rare Book Shop Publishing Company, Inc., 1914.

Johnson, Allen, and Malone, Dumas (eds.). *Dictionary of American Biography.* 20 vols. New York: Charles Scribner's Sons, 1928-1937.

Mathews, Mitford M. (ed.) *Dictionary of Americanisms.* 2 vols. Chicago: University of Chicago Press, 1951.

U. S. Department of Commerce. *Units of Weight and Measure: Definitions and Tables of Equivalents.* (National Bureau of Standards Miscellaneous Publication No. 214.) Washington: Government Printing Office, 1955.

Secondary Accounts: Books

Abernethy, Thomas P. *Western Lands and the American Revolution.* New York: D. Appleton-Century Company, 1937.

Adams, Herbert B. *Maryland's Influence upon Land Cessions to the United States.* (Johns Hopkins University Studies in Historical and Political Science, Series III, No. 1.) Baltimore: Johns Hopkins University Press, 1885.

Andrews, Charles M. *The River Towns of Connecticut.* (Johns Hopkins University Studies in Historical and Political Science, Series VII, Nos. 7, 8, 9.) Baltimore: Johns Hopkins University Press, 1889.

Bancroft, George. *History of the Formation of the Constitution of the United States of America.* 2 vols. New York: D. Appleton and Company, 1882.

Barnhart, John D. *Valley of Democracy.* Bloomington, Indiana: Indiana University Press, 1953.

Barrett, Jay A. *Evolution of the Ordinance of 1787.* New York: G. P. Putnam's Sons, 1891.

Bond, Beverley W., Jr. *The Foundations of Ohio.* Vol. I of *The State of Ohio.* Edited by Carl Wittke. Columbus, Ohio: Archaeological and Historical Society, 1941.

Boesch, Hans. *Amerikanische Landschaft.* (Neujahrsblatt der Naturforschenden Gesellschaft in Zurich auf das Jahr 1955.) Zurich, 1955.

Burnet, Jacob. *Notes on the Early Settlement of the North-Western Territory.* New York: D. Appleton & Co., 1847.

Burnett, Edmund C. *The Continental Congress.* New York: The Macmillan Company, 1941.

The Centennial Celebration of Rutgers College, June 21, 1870, with an Historical Discourse Delivered by Hon. Joseph P. Bradley, and Other Addresses and Proceedings. Albany, New York, 1870.

Conover, Milton. *The General Land Office: Its History, Activities and Organization.* Baltimore: The Johns Hopkins Press, 1923.

Donaldson, Thomas. *The Public Domain, Its History, with Statistics.* (U. S. House Doc. No. 45, Pt. 4, 46th Cong., 3d Sess.) Washington: Government Printing Office, 1884.

Downes, Randolph C. *Council Fires on the Upper Ohio: A Narrative*

of Indian Affairs in the Upper Ohio Valley until 1795.
Pittsburgh: University of Pittsburgh Press, 1940.

──────. Frontier Ohio, 1788-1803. Columbus, Ohio: Ohio State Archaeological and Historical Society, 1935.

Fenneman, Nevin M. Physiography of the Eastern United States. New York: McGraw-Hill Book Company, 1938.

Final Report, Ohio Cooperative Topographic Survey. 4 vols. Press of the Ohio State Reformatory, 1916-1933. Vol. I: C. E. Sherman (ed.), The Ohio-Michigan Boundary. Vol. III: C. E. Sherman, Original Ohio Land Subdivisions. Vol. IV: C. E. Sherman, Miscellaneous Data.

Ford, Amelia C. Colonial Precedents of Our National Land System as It Existed in 1800. (Bulletin of the University of Wisconsin, No. 352.) Madison: University of Wisconsin, 1910.

Ford, Edward. David Rittenhouse, Astronomer-Patriot, 1732-1796. Philadelphia: University of Pennsylvania Press, 1946.

Freeman, Douglas Southall. George Washington: A Biography. 4 vols. New York: Charles Scribner's Sons, 1948-1951. Vol. I.

Garland, John H. (ed.) The North American Midwest: A Regional Geography. New York: John Wiley and Sons, 1955.

Gilmore, William E. Life of Edward Tiffin, First Governor of Ohio. Chillicothe, Ohio: Horney & Son, 1897.

Gipson, Lawrence Henry. Lewis Evans. Philadelphia: The Historical Society of Pennsylvania, 1939.

Hening, William W. (ed.) The Statutes at Large; Being a Collection of All the Laws of Virginia, from the First Session of the Legislature, in the Year 1619. 13 vols. Richmond, Virginia: Printed for the Editor, 1819-1823. Vols. IX, X, XII and XIII.

Hibbard, Benjamin Horace. A History of the Public Land Policies. New York: The Macmillan Company, 1924.

Hildreth, Samuel P. Pioneer History: Being an Account of the First Examinations of the Ohio Valley and the Early Settlement of the Northwest Territory. Cincinnati: H. W. Derby and Co., 1848.

Hinsdale, B. A. The Old Northwest. New York: Silver, Burdett and Company, 1888.

Hosack, David. A Bibliographical Memoir of Hugh Williamson, M.D., LL.D. New York, 1820.

Howard, George E. An Introduction to the Local Constitutional History of the United States. (Johns Hopkins University Studies in Historical and Political Science. Extra Vol. IV.) Baltimore: Johns Hopkins University Press, 1889.

Jacobs, James Ripley. The Beginning of the U.S. Army, 1783-1812. Princeton, New Jersey: Princeton University Press, 1947.

James, James A. The Life of George Rogers Clark. Chicago: University of Chicago Press, 1928.

Johnson, Frank M. "The Rectangular System of Surveying," in U. S. Department of the Interior, General Land Office, Public Land System of the United States. Washington: Government Printing Office, 1924.

Kiely, Edmond R. Surveying Instruments, Their History and Classroom Use. (National Council of Teachers of Mathematics, Nineteenth Yearbook.) New York: Bureau of Publications, Teachers College, Columbia University, 1917.

Legnazzi, E. N. Del Catasto Romano e di alcuni strumenti antichi del geodesia. Padua: Drucker & Tedeschi, 1887.
    Interpretive summary of part referred to in this study supplied by F. J. Marschner, U. S. Department of Agriculture.

Livermore, Shaw. Early American Land Companies: Their Influence on Corporate Development. New York: The Commonwealth Fund 1939.

Jensen, Merrill. The New Nation: A History of the United States during the Confederation, 1781-1789. New York: Alfred A. Knopf, 1950.

Lokken, Roscoe L. Iowa Public Land Disposal. Iowa City: The State Historical Society of Iowa, 1942.

Manley, Henry S. The Treaty of Fort Stanwix, 1784. Rome, New York: Rome Sentinal Company, 1932.

Pence, George, and Armstrong, Nellie C. Indiana Boundaries: Territory, State, and County. Indianapolis: Indiana Historical Bureau, 1933.

Peters, William E. Ohio Lands and Their History. 3d ed. Athens, Ohio: W. E. Peters, 1930.

Phillips, Philip Lee. The First Map and Description of Ohio, 1787, by Manasseh Cutler: A Bibliographical Account. Washington: W. H. Lowdermilk and Co., 1918.

Randall, John H., and Haines, George. "Controlling Assumptions in the Practice of American Historians," Chapter II of Theory and Practice in Historical Study: A Report of the Committee on Historiography. (Social Science Research Council Bulletin 54.) New York: Social Science Research Council, 1946. Pp. 15-52.

Sato, Shosuke. History of the Land Question in the United States. (Johns Hopkins University Studies in Historical and Political Science, Series IV, Nos. 7, 8, 9.) Baltimore: Johns Hopkins University Press, 1886.

Savelle, Max. George Morgan, Colony Builder. New York: Columbia University Press, 1932.

[Smith, William.] An Historical Account of the Expedition against the Ohio Indians, in the Year 1764, under the Command of Henry Bouquet, Esq.; Colonel of Foot, and Now Brigadier General in America. . . . Philadelphia: Printed and Sold by William Bradford, 1765.

Stenton, F. M. Anglo-Saxon England. Oxford: Oxford University Press, 1943.

Stewart, Lowell O. Public Land Surveys: History, Instructions, and Methods. Ames, Iowa: Collegiate Press, Inc., 1935.

Treat, Payson J. The National Land System, 1785-1820. New York: E. B. Treat and Company, 1910.

Williams, Samuel Cole. History of the Lost State of Franklin. New York: The Press of the Pioneers, 1933.

Secondary Accounts: Articles

Alden, George H. "The State of Franklin," *American Historical Review*, VIII (January, 1903), 271-289.

Burt, A. L. "A New Approach to the Problem of Western Posts," *Report of the Annual Meeting of the Canadian Historical Association*. Ottawa, 1931, pp. 61-75.

Beck, T. Romeyn. "Eulogium on the Life and Services of Simeon De Witt," *Transactions of the Albany Institute*, II (1852), 313-315.

Brainard, Newton C. "Colonial Surveying Instruments," *Connecticut Historical Society Bulletin*, XIV (April, 1949), 10-12.

Caterini, Romolo de. "Gromatici Veteres, I tecnici erariali dell'antica Roma," *Revista del castasto e dei servizi tecnici erariali*, II (June, 1935), 261-358.
    Interpretive summary of part of article referred to in this study supplied by F. J. Marschner, U. S. Department of Agriculture.

Culley, John L. "Steel Tapes," *Journal of the Association of Engineering Societies*, VI (August, 1887), 305-310.

Cumrine, Boyd. "The Boundary Controversy between Pennsylvania and Virginia; 1748-1785," *Annals of the Carnegie Museum*, I (1901), 505-524.

Downes, Randolph C. "Ohio's Squatter Governor: William Hogland of Hoglandstown," *Ohio Archaeological and Historical Publications*, XLIII (1934), 274-275.

Dyer, Albion M. "First Ownership of Ohio Lands," *New England Historical and Genealogical Register*, LXIV (April, July, and October, 1910), 167-180, 263-282, 356-369; and LXV (January, April, and June, 1911), 51-62, 139-150, 220-231.

Freund, Rudolph. "Military Bounty Lands and the Origin of the Public Domain," *Agricultural History*, XX (January, 1946), 8-18.

Graham, Louis E. "Fort McIntosh," *Western Pennsylvania Historical Magazine*, XV (January, 1932), 93-119.

Hulbert, Archer B. "The Indian Thoroughfares of Ohio," *Ohio Archaeological and Historical Publications*, VIII (1900), 263-295.

Jensen, Merrill. "The Cession of the National Domain, 1781-1784," *Mississippi Valley Historical Review*, XXVI (December, 1939), 323-342.

_____. "The Cession of the Old Northwest," *Mississippi Valley Historical Review*, XXIII (June, 1936), 27-48.

Johnson, Hildegard Binder. "Rational and Ecological Aspects of the Quarter Section: An Example from Minnesota," *Geographical Review*, XLVII (July, 1957), 330-348.

Kingman, E. D. "Roger Sherman, Colonial Surveyor," *Civil Engineering*, X (August, 1940), 514-515.

Langewiesche, Wolfgang. "The United States from the Air," *Harper's Magazine*, CCI (October, 1950), 176-198.

Moore, H. C. "Origin and Authorship of the Present System of Government Land Surveys," *Journal of the Association of Engineering Societies*, II (July and August, 1883), 282-287.

Pattison, William D. "Use of the U. S. Public Land Survey Plats and Notes as Descriptive Sources," *Professional Geographer*, New Series, VIII (January, 1956), 10-14.

Pease, Theodore C. "The Ordinance of 1787," *Mississippi Valley Historical Review*, XXV (September, 1938), 167-180.

Peffer, E. Louise. "Which Public Domain Do You Mean?" *Agricultural History*, XXIII (April, 1949), 140-146.

Pershing, Benjamin H. "A Surveyor in the Seven Ranges," *Ohio State Archaeological and Historical Quarterly*, XLVI (1937), 257-270.

Porter, William A. "A Sketch of the Life of General Andrew Porter," *Pennsylvania Magazine of History and Biography*, IV (1880), 261-301.

Raup, H. F. "The Names of Ohio Streams," *The Ohio Conservation Bulletin*, XX (July, 1956), 10-11.

Reed, Susan M. "British Cartography of the Mississippi Valley in the Eighteenth Century," *Mississippi Valley Historical Review*, II (September, 1915), 213-224.

Sears, Paul B. "The Natural Vegetation of Ohio, I: A Map of the Virgin Forest," *Ohio Journal of Science*, XXV (May, 1925), 139-149.

Sioussat, St. George L. "The Chevalier de la Luzerne and the Ratification of the Articles of the Confederation," *Pennsylvania Magazine of History and Biography*, LX (October, 1936), 391-418.

_____. "The North Carolina Cession of 1784 in Its Federal Aspects," *Proceedings of the Mississippi Valley Historical Association*, II (1908-1909), 35-62.

Smith, Guy-Harold. "The Relative Relief of Ohio," *Geographical Review*, XXV (April, 1935), 272-284.

_____. "Washington's Camp Sites on the Ohio River," *Ohio Archaeological and Historical Quarterly*, XL (January, 1932), 1-19.

Stout, Wilbur, and Lamb, G. F. "Physiographic Features of Southeast Ohio," *Ohio Journal of Science*, XXXVIII (March, 1938), 1-35.

Tatter, Henry. "State and Federal Land Policy during the Confederation Period," *Agricultural History*, IX (October, 1935), 176-186.

Teetor, Henry B. "Israel Ludlow and the Naming of Cincinnati," *Magazine of Western History*, II (July, 1885), 251-257.

Treat, Payson. "Origin of the National Land System under the Confederation," *American Historical Association Annual Report for the Year 1905*, I (1906), 231-239.

Truesdell, W. A. "Origin of the United States Land Surveys," *Journal of the Association of Engineering Societies*, XXXII (April, 1904), 194-201.

_____. "The Rectangular System of Surveying," *Journal of the Association of Engineering Societies*, XLI (November, 1908), 207-230.

Tunnard, Christopher. "Fire on the Prairie," *Landscape*, II (Spring, 1952), 9-13.

Turner, Frederick Jackson. "The Old West," *Wisconsin State Historical Society Proceedings, 1908* (1909), pp. 184-233.

———. "Western State-Making in the Revolutionary Era," *American Historical Review*, I (October, 1895), 70-78, and I (January, 1896), 251-269.

Tyrrell, George F. "Background and Development of Cadastral Surveys," *Surveying and Mapping*, XVII (January, February, March, 1957), 33-41.

Verner, Coolie. "The Maps and Plates Appearing with the Several Editions of Mr. Jefferson's 'Notes on the State of Virginia,'" *Virginia Magazine of History and Biography*, LIX (January, 1951), 21-33.

Whittlesey, Charles. "Origin of the American System of Land Surveys," *Journal of the Association of Engineering Societies*, III (September, 1884), 275-280.

Wilson, George R. "Early Indiana Trails and Surveys," *Indiana Historical Society Publications*, VI (1919), 347-457.

Newspaper

New York Times, May 19, 1957.

Miscellaneous Unpublished Materials

Hutchinson, William T. "The Bounty Lands of the American Revolution in Ohio." Unpublished Ph.D. dissertation, Department of History, University of Chicago, 1927.

Lehmann, Herbert. "The Role of Law and Tradition in the Use of Agricultural Resources." Paper in Report of Seminar on Agricultural Utilization of Natural Resources, University of Chicago, Spring and Summer Quarters, 1952. Chicago: Department of Geography, University of Chicago, November, 1952. (Mimeographed.)

Memorandum from F. J. Marschner, U. S. Department of Agriculture. April 15, 1956.

Pershing, Benjamin H. "Winthrop Sargent: A Builder in the Old Northwest. Unpublished Ph.D. dissertation, Department of History, University of Chicago, 1927.

Quattrocchi, Anna Margaret. "Thomas Hutchins, 1730-1789." Unpublished Ph.D. dissertation, Department of History, University of Pittsburgh, 1944.

Smith, Dwight. "Indian Land Cessions in the Old Northwest, 1785-1809." Unpublished Ph.D. dissertation, Department of History, Indiana University, 1949.

Maps and Atlases

Gannett, Henry (ed.). *Statistical Atlas of the United States*. U. S. Department of the Interior, Census Office. Washington: Government Printing Office, 1898.

Hough, B., and Bourne, A. *Map of the State of Ohio from Actual Survey*. Philadelphia: B. Hough & A. Bourne and J. Melish, 1815. Scale: 1 inch to about 5 miles.

Hutchins, Thomas. A New Map of the Western Parts of Virginia, Pennsylvania, Maryland and North Carolina: Comprehending the River Ohio and All the Rivers, Which Fall into It; Part of the River Mississippi, the Whole of the Illinois River, Lake Erie; Part of Lakes Huron, Michigan &c. And All the Country Bordering on These Lakes and Rivers. London: Engraved by I. Cheevers, 1778. Scale: 1 inch to about 20 miles.

Mansfield, John F. Map of the State of Ohio Taken from the Returns in the Office of the Surveyor General. 1806. Scale: 1 inch to about 10 miles.

"A Map of the Country between Albermarle Sound, and Lake Erie, Comprehending the Whole of Virginia, Maryland, Delaware, and Pennsylvania, with Parts of Several Other of the United States of America," in Paul Leicester Ford (ed.), The Writings of Thomas Jefferson. 10 vols. New York: G. P. Putnam's Sons, 1892-1899. Vol. III, foll. p. 84. Scale: 1 inch to about 20 miles.

"Map of the District of Maine," in James Sullivan, The History of the District of Maine. Boston, 1795. Frontispiece. Scale: 1 inch to about 10 miles.

"A Map of the Federal Territory from the Western Boundary of Pennsylvania to the Scioto River Laid Down from the Latest Informations . . . ," in Phillips, The First Map of Ohio, foll. p. 41. Scale: 1 inch to about 10 miles.

"Map of Ohio Showing Original Land Subdivisions," Accompanying Final Report, Ohio Cooperative Topographic Survey, Vol. III. Scale: 1 inch to 6 miles.

"Map of the State of Ohio by Rufus E. Putnam, Surveyor General of the United States," in Thaddeus M. Harris, The Journal of a Tour into the Territory Northwest of the Alleghany Mountains; Made in the Spring of the Year 1803. Boston, 1805. Foll. p. 271. Scale: 1 inch to 20 miles.

Nederlanden Topographische Dienst. Chromo-Topographische Kart des Rikjs. Scale: 1:25,000. Sheet No. 280 (1906) and Sheet No. 296 (1907).

Paullin, Charles O. Atlas of the Historical Geography of the United States. Edited by John K. Wright. (Carnegie Institution of Washington Publication No. 401.) Washington: Carnegie Institution of Washington, 1932.

"A Plan of the Several Villages in the Illinois Country," in Hutchins, A Topographical Description (1904), facing p. 112. Scale: 1 inch to about 10 miles.

Plat of the Seven Ranges of Townships Being Part of the Territory of the United States N.W. of the River Ohio. [Pub. by Mathew Carey, 1796.] Scale: 1 inch to 4 miles.

"Plat of the Seven Ranges of Townships, Ohio Survey, 1785-87," in Elroy M. Avery, A History of the United States and Its People. 7 vols. Cleveland: The Burrows Brothers, 1904-1910. Vol. VI, foll. p. 406. Scale: 1 inch to about 8 miles.

U. S. Department of the Interior, General Land Office. State of Illinois. Washington: Government Printing Office, 1911. Scale: 1 inch to 12 miles.

─────── . State of Indiana. Washington: Government Printing Office, 1916. Scale: 1 inch to 12 miles.

U. S. Geological Survey. Topographic Map of the United States.
Washington: Government Printing Office. Scale: 1:62,500.
Selected Ohio Sheets (1904-1942).

"United States Development to 1787," in Elroy M. Avery, A History
of the United States and Its People. 7 vols. Cleveland:
The Burrows Brothers, 1904-1910. Vol. VI, foll. p. 410.
Scale: 1 inch to about 75 miles.

INDEX

ACRES, land measure, 49, 56, 94.
Adams, John, 31n.
Alabama, boundary, 28-29.
Alexander, Major _____, of Virginia, 201n.
Allegheny Front, 108, 117.
Allegheny Plateau, 108, 116, 117, 162.
Amplitudes, 79.
Appalachian Mountains, 28n, 108, 116.
"Army Plan," for western settlement, 23, 41.
Articles of Confederation, 6n, 7, 72.
Auctions, see Public land auctions.
Axe-men, 121, 152, 202, 209.

BACKUS, Elijah, 200n.
Base lines, in surveying, 53, 182, 210-212; in Indiana, 222; suggested by Jefferson, 226.
Beginning point of survey, 102-103, 114; established, 119-122, 127.
Bever, John, 207n.
Big Miami River, see Great Miami River.
Biggs, Zaccheus, 202n, 207n, 208n.
Biram River, 6n.
Board of Treasury, administration of public lands by, 100n, 101, 138, 139-140, 143, 149, 150, 155n, 157, 160, 162n, 164, 165n; and numbering system, 98, 197; contract with Ohio Company, 169, 178n; and Symmes Purchase, 170; and Hutchins, 173, 175.
Boudinot, Elias, opposed indiscriminate location, 186.
Bouquet, Colonel Henry, 58-59, 61-62, 65.
British, at Detroit, and Indians, 113-114.
British posts in Northwest, surrender of, 188.
Buckingham, Ebenezer, 207n.
Bull, John, 85n.
Bureau of Land Management, 207n.
Butler, General Richard, 14n, 132n.

CAHOKIA (Illinois), 6n.
Carey, Matthew, map of the Seven Ranges, 165, 218, 219, 228.
Carlisle, Pennsylvania, 107.
Central Lowland, 118.
Cessions, see Land cessions.
Chain carriers, 152, 198, 208, 209.
Chains, surveyors', 73, 74-75, 79, 81, 97, 104, 148, 214, 224; Jefferson's proposed, 49; specified by Land Act of 1796, 196.
Chase, Jeremiah T., member Jefferson's committee on western lands, 15n.
Chattahoochee River, 29n.
Chicago, Illinois, site of, surrendered by Indians, 190n.
Chillicothe, Ohio, land office, 205n.
Chippewa Indians, at Fort McIntosh, 14; treaty with, 15n; land ceded by Treaty of Greenville, 188, 190.
Cincinnati, Ohio, 188; land office, 205n.
Circumferentors, use in surveying, 77-78, 146, 148, 215, 224.
College of William and Mary, 72.
Committee on Indian affairs, report of October 15, 1783, 11; membership, 13.
Committee on public lands (Jefferson's, 1784), 4, 11, 63; membership, 3; report of, 17, 56, 66, 77, 79, 80, 81, 82, 88, 90, 94, 95, 97, 100-101, 104; proposed ordinance, 37, 84n.
Compass, magnetic, 73, 77-79, 104, 145, 148, 215; solar, 215.
Connecticut, western land claim, 5-6, 10, 20; land cession, 8; prior survey system in, 87; western surveyor for, 125.
Connecticut Western Reserve, 8, 125, 126, 199; title confirmed, 132n; township size in, 194.

249

Constitution of the United States, enacted, 185.
Continental Congress, 3, 37, 191; termination, 185n.
Contract system, 179-180, 198, 208-209; terminated, 209n; origin, 225.
Convergency problem, 51-56, 66, 88-89, 97, 210, 212; recognition of, prior to 1800, 222-223.
Cox's Fort, 111.
Crabb, Jeremiah, 196.
Creeks, named in early survey notes, 168n.
Cumberland Valley, 107-108.
Cutler, Manasseh, map, 164-165, 228.
Cuyahoga River, as Indian boundary (text has Cayahoga), 14, 190n.

DANE, Nathan, opposed rectangular grid, 89.
Dawson farm, Ohio River settlement, 111.
Delaware (state), adopted division into hundreds, 43.
Delaware Indians, 114, 115, 116, 129, 132; and Treaty of Fort McIntosh, 14; chiefs refused surveyors safe conduct, 135; hostility, 133n, 187; reservations for, 175, 199n, 203n; land ceded by Treaty of Greenville, 188, 190.
Detroit, Michigan Territory, British outpost, 113-114, 115.
De Witt, Simeon, 60, 61, 65, 123n; surveyor general of New York, 69-70; resignation as Geographer, 100; declined Surveyor General's post, 200.
Dexter, Samuel, 206n.
Donation Tract, 178.
Dowse, Edward, 124, 133.

EAST AND WEST LINE, see Geographer's Line.
Edgar, William, 156n.
Eel River Indians, 190n.
Ellicott, Andrew, 121, 122, 123n, 176n; and survey of New York boundary, 173.
Ellicott's Line, 122, 128.
England, land system as model, 43, 59n.
Expense of surveying, 149-154, 224-225.

FAIRFAX, George William, 70.
Fallen Timbers, Battle of, 15, 30, 188, 225.
Falls of the Ohio (Rapids of the Ohio), 13, 19n, 25; meridian at, 17, 25, 26n, 28n.
Federal troops, and protection of surveyors, 129-130, 135-136, 137, 141, 142; retreated from field, 138.
Field books, 196. See also Survey notes.
Figures, Jefferson's land subdivisions, 47; solution to problem of rectangles and meridians, 54. See also Maps and diagrams.
"Financier's Plan," for western settlement, 20, 41.
Flint, Abel, 79n.
Flint, Royal, 191n.
Flushing Escarpment, 117.
Forbes Road, 108.
Forests in Seven Ranges, indicated by surveyors' notes, 166.
Fort Greenville, General Wayne's headquarters, 188.
Fort Harmar, 138, 142, 173, 174; established, 131.
Fort McIntosh, 105, 109-110, 116; troops from, failed to support survey party, 129-130.
Fort Recovery, 190n, 202.
Fort Steuben, 141.
Fort Wayne, Indiana, site of, occupied by Wayne's army, 188; site of, surrendered by Indians, 190n.
Founders of American survey system, 220, 225-227.
Franklin, Benjamin, 31n, 123n.
Franklin, state of, projected, 26n, 28n.
Frauds in land sales, 40-41.
French grant (Gallipolis), survey of, 200, 207n.
French villages (Illinois country), 6.

GALLATIN, Albert, 206; and state boundaries, 213, 214; as founder American survey system, 226; as Secretary of the Treasury, responsible for administration of surveying, 227.

Gardner, Joseph, 85n.
Gardoqui, Diego de, 174, 175, 176.
General Land Office, 207, 217n; proposed, 186; established, 206; commissioner of, 206.
Genn, James, 70.
Geographer of the United States, 105; duties, 100-101, 160; salary, 153, 154; post left vacant, 176. See also De Witt, Simeon, and Hutchins, Thomas.
Geographer's Department, 123, 150.
Geographer's Line, 144, 160n, 170, 199, 203, 213, 215; error in, 146.
Geographical mile, 46-50, 66, 220n.
Georgia, proposed boundary, 25; western land cession, 28, 29n; land claim, 29; survey agent, 128; voided western land sale, 191n.
Gerry, Elbridge, 87n; member committee on western lands, 3; attitude on Indian cessions, 15n.
Gnadenhutten, Ohio, 203n.
Gorges, Sir Ferdinando, 44n.
"Grand Committee" (1785), 84-85, 86, 87. See also Grayson's committee.
Grayson, William, 82-83, 84, 85; boundary plan, 33, 35; quoted, 84, 86-87, 92-93; urged rectangular survey as less expensive, 88, 230; as founder American survey system, 226.
Grayson's committee, report of, 83, 92, 93, 95, 100.
Great Britain, treaty of peace with (1784), 10-11, 20; Jay's Treaty (1794), 188.
Great Lakes, 11
Great Miami River, 11, 14n, 15n, 29, 33, 35, 170, 181, 182, 190n, 202, 203, 204.
Great Smokey Mountains, 24, 28n.
Great Valley, 107-108.
Greathouse, William, 111.
Greenville Treaty Line, 199, 203, 204; demarked, 199, 201-202.
Grid, see Rectilinear grid.
Gunter, Edmund, 75n.
Gunter's chain, 49, 74-75, 148, 196n, 214. See also Chains, surveyors'.

HAMILTON, Alexander, recommendations for sale of western lands, 186; 1790 report of, 194, 198.
Hardinsburg, Kentucky, 26n.
Hardy, Samuel, 84n.
Harmar, Colonel Josiah, 109, 110, 134, 141-142.
Harris, Caleb, 124n, 134n.
Hartley, David, map of proposed states, 25, 26.
Havens, Jonathan, 193, 194; as founder American survey system, 226.
Henry, John, 85n.
Hogendorp, G. K. van, 62.
Hoglands Town, 113.
Holland, influence on American survey system, 62, 63.
Hoops, Adam, 124n, 135, 136, 139n, 140; survey agent for Pennsylvania, 134.
Hough and Bourne map of Ohio, 219.
Houston, William, 85n.
Howell, David, 84n; member Jefferson's committee on western lands, 3, 4, 15n, 85; on committee on Indian affairs, 13; quoted, 25, 39, 82; and land sale, 94.
Human habitation sites in Seven Ranges, 167, 229.
Hundreds, 40, 42, 66; proposal for, 37-38, 43, 46; Jefferson urged use of, 44-46, 63; compared with townships, 45; subdivision of, 56, 70; numbering of, 80, 81.
Hutchins, Thomas, 100, 101, 113, 114, 127, 155, 157, 160, 164, 165n, 169, 177, 181n, 182, 199, 200, 201, 203, 209, 213; map of the West, 17, 19, 26, 123, 218; accuracy in locating latitudes, 20; and convergency problem, 52; claim as founder American system disproved, 65; quoted on office of register, 69; geographer under Land Ordinance of 1785, 70; consultation with Colonel Harmar, 110; at Pittsburgh and Fort McIntosh, 122; sketch of, 123n; and fear of Indians, 128, 129; survey of base

line, 128; retreat from field of survey, 130, 136; description
of country along first survey line, 130-131; resumption of surveying, 132-136, 137; preparation of plats, 139; departure from
West, 139; submission of notes and plats to Board of Treasury,
140, 143; and running of East and West Line, 144-146; remuner-
ation, 149-151; at land auction, 156; testimonial used by Ohio
Company, 162; last days, 172-176; death, 172, 176; despair,
175-176, 225; as developer survey system, 227; preference for
natural boundaries, 230.

ILLINOIS, present northern boundary, 34, 35; shatter zone in, 55.
Illinois and Wabash Company, 191n.
Illinois country, 6, 10, 176; application for grants in, 191.
Illinois Indian villages, 6n.
Illinois River, 32.
Indiana, boundaries prescribed by Ordinance of 1787, 34, 35; corrective parallel in, 55; surveys in, 55; state boundary, 213;
base line and principal meridian in, 222.
Indians, Northwestern tribes, land cessions by, 4, 10-15, 30, 109,
114, 188, 190 boundaries for, 11, 13, 15n, 25, 29, 30-31, 190;
Northwestern confederacy (1783), 13, 14; treaties with, 13-14,
15, 82, 114, 116, 129, 130n, 177, 188, 190; raids into ceded
lands, 15; wars with, 15, 115, 137, 187-188; Washington quoted
on peace with, 29; beyond Seven Ranges, 105; in Seven Ranges,
113-116; threaten resistance to surveyors, 128, 129, 137, 141,
142; 175-176; introduced to survey methods, 163; land cessions
sought from, 170; peace negotiations, 187; defeated in Northwest, 187-190; and Greenville Treaty Line, 202; delayed surveying in Seven Langes, 224.
Interior Department, see United States Department of the Interior.
Iroquois Indians, see Six Nations.

JACKSON, John, 202n, 207n.
Jacob's staff, 78.
Jay's Treaty, 188.
Jefferson, Thomas, chairman, committee on western lands, 3; committee on Indian affairs, 13; plan for western states, 15-30,
127, 146, 194, 197, 213; on Indian cessions, 15n; estimate of
Hugh Williamson, 38; as founder rectangular survey system, 38,
66, 225, 226; urged use of decimal system, 43-50, 57; urged
use of hundreds, 44-46; and geographical mile, 46-50; aware of
convergency problem, 51; on orientation to compass, 63; borrowed register idea, 68-69; and Virginia land laws, 72, 76, 77;
view of Ordinance of 1785, 103-104; quoted, 121; used Hutchins'
map, 123n; influence on Ordinance of 1785, 132; appointment to
office by, 201; removal from office by (Putnam's), 207. See
also Committee on public lands.
Jefferson-Williamson plan, 38-39, 40, 42, 53, 56, 57-58, 59, 60,
63, 64, 68, 86, 88, 89, 92, 93, 163, 183, 210, 220n, 226, 230.
Johnson, William Samuel, 84n, 85n.
Johnston, Robert, 128, 133n.

KANAWHA MERIDIAN, 17, 22, 23, 24-25, 27, 28n, 30.
Kanawha River, 17, 22, 24n, 25.
Kanawha Valley, 24.
Kaskaskia, Illinois, 6n.
Kaskaskia Indians, 190n.
Kekewepellethy, Miami chief, 115n.
Kentucky, land open to sale, 7; set off from Virginia, 23-24; bound-
aries, 26, 28; war with Indians, 115; militia activities, 137;
volunteers from, in Wayne's army, 188.
Kentucky Military District, 10, 42.
Kentucky River, 130n, 202.
King, Rufus, quoted, 83, 85, 87, 88, 93; and Land Ordinance of 1785,
85, 87, 90, 92, 95, 132; and survey of Virginia Military District,
95; as founder American survey system, 226.
Kittera, 197n.

LAKE ERIE, 32, as boundary, 8, 11, 14, 17, 22, 29, 104, 199.
Lake Michigan, 32; as boundary, 26, 33, 35.

Lake Ontario, 21.
Land Act of 1796, 2, 53, 92, 95, 102, 119, 179, 180, 187, 214, 216, 217, 224; and term "section," 57; federal surveying under, 185-204; areas where applicable, 189, 190, 199; approved by George Washington, 192; legal basis of federal surveying, 193; surveying and numbering under, 196-198, 210; surveyors' duties under, 198-199; field notes under, 203n, 216; amendment, 205; and convergency problem, 210; townships under, 210, 221; quality of surveying under, 214-216; lot closure under, 221-222; Putnam as Surveyor General under, 223; perpetuated rectangular survey, 226; advocates of, 226; development of surveying under, 227; provision for strict boundaries, 230.
Land Act of 1800, 95n.
Land cessions, state, 3, 4-10, 19, 20, 24-25, 28, 29n, 41; Indian, 4, 10-15, 30, 109, 188, 190.
Land companies, lost leadership, 190-192. See also names of companies.
Land evaluation, 168, 229.
Land law of 1784 (proposed), 81; required grid, 220; state boundaries proposed by, 222.
Land law of 1805, 205n.
Land offices, 84n, 205n; proposal for, 37; under law of 1800, total sales, 205n; plats prepared for, 218.
Land Ordinance of 1785, 1, 37, 51, 57, 65, 70, 76, 77, 79, 81, 82-104, 139, 155, 158, 178, 182, 213, 227; passage, 83; summarized, 85-86; proposed, 88; motion to amend, 90; reserved lands under, 95-96; surveyors under, 101-102, 122-128, 223; federals surveying under, 105-143; private surveys under, 119, 169-184; fixed rates of pay for Geographer and surveyors, 153n; supplement to, 157n; foundation of rectangular survey system, 159; expiration, 185, 186; and Land Act of 1796, 193, 194, 196, 198, 201, 214; exempted surveyors from orienting to true north, 224; Grayson's role in framing, 226.
Land sales, 192n; under Ordinance of 1785, 153-157; under 1800 law, 205. See also Public land auctions.
Land warrant system, 40, 42, 68n, 69n, 70-71, 101, 183; abandoned, 100.
Langham, Elias, 207n.
Lee, Arthur, 123n; quoted, 107.
Lee, Richard Henry, 14n, 83-84.
Lehigh Valley (Pennsylvania), 6n.
Little Beaver Creek, 123, 127, 128.
Little Miami River, 170, 180, 181, 182, 203n.
Livingston, Robert R., 85n.
Logan, Colonel Benjamin, 137.
Long, Pierse, 85n.
Long Island Sound, 6n.
Loramies Store, 202n.
Lots, identification of, 79-81, 98, 99, 104; term for "sections," 94; sale of, 104.
Ludlow, Israel, 140, 141, 143n, 158n, 159n, 160n, 178, 203, 223; survey agent for South Carolina, 134; and survey of Seven Ranges, 134, 140-142, 143n, 178, 223n; surveyor for Ohio Company, 174, 177, 223n; surveyor of Symmes Purchase, 181-184, 223n; applied for Surveyor General's post, 200n; surveying under Land Act of 1796, 201-204, 223n; work done under Putnam, 201-202, 207n; as land office official, 208n; compensation, 209n; established principal meridian, 212, 213.

McCALL, Mark, 124n.
McComb, Alexander, 156n.
McLean, Archibald, 122n.
McMahon, William, home of, used by surveyors, 111, 136, 138, 142, 143.
Madison, James, 24.
Magnetic variation, 78, 79.
Maine, hundreds suggested for, 44n; dishonest surveys in, 71; Putnam's survey in, 90.
Mansfield, Jared, 145n, 215, 222n; as developer American survey system, 65, 66, 227; as Surveyor General, 201n, 207, 226; and

solution of surveying problems, 210, 212; established master
    survey lines, 210, 222n; and state boundaries, 213-214; closure
    plan used by, 216.
Mansfield, John P., map of Ohio, 219.
Map making, aided by survey records, 228.
Maps, of West (Hutchins'), 17-19, 26, 123, 218; of proposed states
    (Hartley's), 25 26; of Seven Ranges (Carey's), 165, 218, 228;
    of eastern Ohio (Cutler's), 164-165, 228; of Ohio (Putnam's),
    218-219, 228; of Ohio (John P. Mansfield's), 219.
Maps and diagrams, rectangular survey system, extent of, Frontis-
    piece; claims in Northwest by Virginia, Connecticut, and Mass-
    achusetts, 9; lands reserved by Virginia, 9; boundaries of
    Indian cessions in the Northwest, 12; projected boundaries for
    western states, 18, 27; state boundaries prescribed by Ordi-
    nance of 1787, 24; solution to problems of parallels and
    meridians, 54; township bounding in New England, 91; number-
    ing of townships and lots under Land Ordinance of 1785, 99;
    first scene of survey and environs, 106; Seven Ranges, 120;
    surveying diagrams, 147; Ohio Company lands and the Miami Pur-
    chase, 171; areas first surveyed under Land Act of 1796, 189;
    details of surveying under act of 1796, 211.
Marietta, 134; founded, 163, 173; platted, 178, 179; land office,
    205n.
Martin, Absalom, 126, 133n, 135, 138, 140, 142, 143, 159n, 167n,
    202n, 207n, 208n, 223n; completed Fifth Range, 142; surveyor
    for Ohio Company lands, 177.
Maryland, refusal to ratify Articles of Confederation, 7; boundary,
    21; divided into hundreds, 43; survey agent for, 127.
Mason and Dixon Line, 5n, 20-21, 119n, 145n.
Massachusetts, western land claim, 5, 10, 140n; boundary, 6n;
    land cession, 8, 33n; prior survey in, 87; survey agent for,
    124.
Massachusetts Bay Company, 5.
Mathews, John, 139n, 159n, 161n, 179n, 202n, 207n, 223n.
Maumee River, 11, 14, 29.
Meridians, as survey lines for bounding, 5, 25, 26, 33, 35, 59,
    144-145; Kanawha, 17, 22, 23, 24-25, 27, 28n, 30; problem of
    convergence, 51-56, 66, 88-89, 97, 210, 212, 222-223; "true"
    required in 1785, 51, 90, 96, 132; principal, 53, 210-212, 222;
    magnetical, 77; "true" not required, 146, 148, 157, 158, 215;
    "standard," 182; "true" required in 1796, 196-197; use of,
    suggested by Jefferson, 226.
Miami fort, 11n.
Miami Indians, 115-116; expeditions against, 187-188; land cessions,
    190.
Miami Purchase (Symmes Purchase), 170, 172, 174, 199, 205n; Ludlow
    interested in, 154; survey of, 180-184, 223n; numbering of town-
    ships in, 184n; map of, 228.
Miami Valley, 187.
Michigan, wilderness in 1818, 35.
Military bounty lands, 100n, 172, 199, 202; survey of, 207n. See
    also Kentucky Military District, Virginia Military District,
    and United States Military Reserve.
Mill sites, located in field notes, 97.
Mines, located in field notes, 97.
Mingo Bottom, 112.
Mississippi River, 32; as boundary, 5, 7, 11, 28, 170, 190, 191n.
Mohawk Valley, 14.
Monroe, James, 84n, 100; ideas on western state-making, 31-33,
    35; quoted on Land Ordinance, 85-86.
Montgomery, Robert, 58, 59n.
Montgomery, Samuel, 133n; quoted, 116n.
Moravian (Delaware Indian) reservations, 175, 199n, 203n.
Morgan, George, 176, 191n.
Morris, William W., 126-127, 133n, 135, 176n.
Muskingum River and Valley, 15n, 112, 114, 117, 133, 134, 142,
    161, 163, 190n, 202n, 205n.

NASHVILLE BASIN, 26.
National domain, 3, 8, 10, defined, 3n; land titles in, 231.

Nautical mile, 46, 63. See also Geographical mile.
Neville, Joseph, 200n.
New England land system, 39, 40, 42, 59, 65, 68n, 80; as model for national surveys, 43n, 44n, 61, 80, 89, 90, 92, 93, 193, 221, 226.
New Hampshire, survey agent, 124, 133.
New Jersey, survey agent, 126, 133.
New York, western land claims, 5; negotiations with Indians, 13n; boundary, 21, 140n; use of rectangular surveys, 61; prior survey in, 87; artillery company at Fort McIntosh, 110, survey agent, 126; boundary survey, 172, 173n; land policy model for federal policy, 193, 226.
Newburgh Petition, 22-23, 125n.
Niagara River, 21n.
Nickajack, Tennessee, 29n.
Norris, Charles, 112.
"Norris Town," 112.
North Carolina, western land claims, 20, 28; land cessions, 24-25; boundary, 28n; survey system, 41; survey law of 1777, 51; land laws as model for national survey system, 58; survey agent, 124n, 133.
Northwest Ordinance, see Ordinance of 1787.
Northwest Territory, see Territory North West of the River Ohio.

OHIO, present state, 11; admission to Union, 31; present northern boundary, 34, 35; convergence of survey lines in, 55; early maps of, 164-165, 218-219, 228.
Ohio Company of Associates, 125, 133, 183, 199, 205n; beneficiary of survey of Seven Ranges, 160-163, 225; survey of lands of, 160-163, 172, 175, 177-180; land grant to, 162; negotiations with Congress, 169; first settlement, 173; Hutchins as surveyor for, 173-174; boundary survey, 181n; Superintendent of Surveys for, 179, 223n; Indian threat to, 187; failure to fulfill contract, 191; map of lands of, 228.
Ohio country, map of, 228.
Ohio River, 107, 109, 112, 113, 115, 117, 118, 119, 122, 123, 130, 134, 136, 139, 141, 142, 143, 156, 157, 175, 191, 199, 202; as boundary, 4, 5, 6, 11n, 13, 14, 15n, 17, 22, 23, 24n, 25, 26, 28, 29, 33, 35, 42, 98, 104, 170, 182, 202; settlements on, 105, 110-113; legal frontier, 110; tributaries, 117, 118; surveyors winter on, 151; boundary of Seven Ranges, 164; charted, 165, 174, 175, 177, 181.
Ordinance of 1784, 15-19, 83, 88, 104n, 169; conditions for, 4; Washington's objections to, 30; Monroe's objections to, 31-32; reappraisal of, 32, 33; and state boundaries, 32, 36, 53, 213, 214; proposal for, 39, 49.
Ordinance of 1785, see Land Ordinance of 1785.
Ordinance of 1787 (Northwest Ordinance), 29; compared with Ordinance of 1784, 16; governmental provisions, 16; basis for, 33; boundaries prescribed by, 33, 35-36; passage, 169; reenacted, 185.
Orientation to compass, 224; Jefferson's idea on, 63, 64.
Ottawa Indians, and Treaty of Fort McIntosh, 14; land cessions, 188, 190.

PACKHORSEMEN, 152.
Paine, Thomas, 25.
Parallels of latitude, "standard," 53; as guide lines in surveying, 55; for bounding, 59.
Parker, Alexander, 127, 133n.
Parsons, Samuel Holden, 125; petitioner for Ohio Company lands, 169n.
Pennsylvania, western land claim, 5, 6n; western boundary survey, 5, 21, 29n, 119, 122, 170, 213; southern boundary, 21n, 22n, 103, 145, 146, 163, 163n-164n; warrant system, 68n; land system, 68n, 230n; militia at Fort McIntosh, 109-110; survey agent, 133, 134.
Pennsylvania Road, 107, 130.
Philadelphia, land sale at, 192n.
Piankashaw Indians, 190n.

Pickering, Timothy, 30, 87, 93, 184n; and "Army Plan," 23; and convergence of meridians, 52, 53, 55; quoted on surveyors' duties, 97.
Pipe, Captain, Wyandott chief, 115n, 129, 130n.
Pittsburgh, and road from East, 105, 107-109; described in 1785, 107; land sale at, 192n.
Place names, indicated in survey notes, 167-168, 229.
Plats, 73, 75-77, 78, 81, 97, 101, 104, 160, 196, 218; preparation of, 139; convention for, 159; historical value, 165-168; scale, 217; duplicate copies required, 217.
Plots, see Plats.
Polder Beester, rectangular surveys in, 62.
Porter, Andrew, 122, 176n.
Potawatomi Indians, 190n.
Prairie du Rocher, 6n.
Price of land, minimum, under Land Ordinance, 86, 156; average bid, 156; under Land Act of 1796, 196.
Prior survey system, 39-40, 93; principle of, in Land Ordinance of 1785, 86-87, 104, 221-222; established by Land Act of 1796, 193; chief advocate, 226.
Public debt, land sales and the, 155.
Public domain, see National domain.
Public land auctions, 86, 87, 93, 100n, 104, 155-156, 160, 193, 196, 221.
Putnam, Israel, 209n.
Putnam, Rufus, 23, 123n, 124-125, 161n, 176n, 213; boundaries proposed by, 30; as developer survey system, 65, 227; and Maine survey, 90; Superintendent of Surveys for Ohio Company, 179-180, 208, 223n; and contract system, 179-180, 198-199, 208-209, 225; contract with Ludlow, 180; as Surveyor General, 200-204, 207; contracts for subdivision of townships, 205; contracts with deputies, 208-209; contribution to public land surveying, 208; and numbering of townships, 212; aware of convergency problem, 212; introduced standard chain, 214; urged repeal of true meridian clause, 215; closure procedure, 216; map of Ohio, 218-219, 228; failed to enforce true north, 221n.
Putnam, William R., 207n.

QUADRANTS, 98, 182.

RANGES, numbering of, 98, 99.
Randall, Robert, land speculation of, 192.
Rapids of the Ohio, see Falls of the Ohio.
Read, Jacob, member committee on western lands, 3.
Record-keeping, early precedent, 158-159, 217, 224.
Rectangular survey system, subdivision advantages, 50-51; origin, 57-66.
Rectilinear grid, 88-92, 220-223; under Land Act of 1796, 193.
Register of the Land Office, as proposed in 1784, 68-70; duties, 72, 76, 81.
Reserved lands, 8-10, 95-96; for military service, 8, 10, 11, 42, 95, 169, 170, 194, 199, 202, 203, 205n; for support of education and religion, 92, 96; salt springs and salt licks, 93, 96n; mineral reservations, 93, 95; Delaware Indians, 175n, 199n, 203n.
Reserves, military, 8, 10, 11, 42, 95, 169, 170, 194, 199, 202, 203, 205n.
Revenue, from sale of public domain, 156-157.
Rhode Island, survey agent, 133-134.
Rittenhouse, David, 119, 121, 128.
Roman surveys, as model for American survey system, 60-62, 63-64.
Ross, Joseph, 112.

SACKETT, Nathaniel, 169n.
St. Clair, Arthur, 123n, 170n; defeated by Indians, 187-188; peace negotiations, 187; preference for natural boundaries, 230.
St. Phillipe, Illinois, 6n.
Salem, Ohio, survey of, 203n.
Salt licks, 93, 96n, 97.
Salt River, 26n.
Salt springs, 93, 96n, 97.

Sandusky, Ohio, 14n.
Sandusky River and Valley, 14n, 115.
Sandusky Trail, 115.
Sargent, Winthrop, 136, 137-138; quoted, 107, 108, 113; survey agent for New Hampshire, 133; reconnoitred for Ohio Company, 161, 208n; applicant for Geographer's post, 177n.
Schoenbrunn, Ohio, survey of, 203n.
Scioto Associates, 178n.
Scioto River, 10, 23, 178, 202, 203; charted, 174, 177.
Scott, Thomas, 185-186.
Secant method, 145n.
Secretary of the Treasury, and administration of public lands, 181, 186, 206, 213, 214, 217n, 218n, 227.
Sections, 56-57, 94-95, 182, 221; sale of, authorized, 194-195; numbering of, under Land Act of 1796, 197; appearance in Land Ordinance, 220.
Seneca Indians, village, 112.
Seven Ranges, 1, 105, 175, 199, 205n, 223; plats of, 98n; beginning point of survey, 102-103; Washington's reference to, 107; lay of land, 116-118; survey of, 119-143, 146-149, 157-159, 172; map of, 120, 165, 218, 228; field notes for, 138, 142-143, 159, 165-168; surveying errors in, 148-149; sale of land in, 155-157; influence of survey of, on later public land surveying, 155, 157-159, 179, 216, 217; benefits of survey, 163-164; land sales reopened in 1796, 192; surveying methods in, 196, 197, 214; section boundaries, 221; survey delayed by Indians, 224.
Sextant, use of, 145, 146.
Shawnee Indians, treaties with, 15n, 130, 132, 190; hostility of, 136, 187; expedition against, 137; land cession, 190.
Shenandoah Valley, 107-108; settlement of, 70.
Sherman, Isaac, 15n, 125-126, 133n, 135, 136, 138, 139.
Sherman, Roger, 125.
Simpson, James, 127, 133n, 135, 139n, 141, 143n, 160n; completed survey of Sixth Range, 142.
Six Nations (Iroquois Indians), leader Northwestern Indian confederacy, 13; and Treaty of Fort Stanwix, 13-14; and Treaty of Fort Harmar, 15n.
Smith, Charles, 133n.
South Carolina, land claim, 20; proposed boundary, 25; survey agent, 133, 134.
Southern land system, 43, 66, 93; influence on federal system, 39-43, 81, 229; abandoned, 86.
Spaight, Richard D., quoted on state cessions, 15n; opposed rectilinear grid, 89.
Sproat, Ebenezer, 135, 136, 138, 161n, 179n; survey agent for Rhode Island, 133-134.
Squatters, in Ohio country, 110, 111-112, 163; in Seven Ranges, 167.
States, land cessions, 3, 4-10, 19, 20, 24-25, 28, 29n, 32-33, 41; western claims, 5-7, 10, 20, 28, 29, 140n; boundaries under Ordinance of 1784, 16-17, 26-29; size of, prescribed in 1780, 19-20; meridians as boundaries, 20-21.
Steubenville, Ohio, land office, 205n.
Stewart, Archibald, 85n.
Superintendent of Surveys for the Ohio Company, 179-180, 208, 223n.
Survey notes, 138, 142-143, 159, 160, 165, 203n, 204n, 216, 217, 228; practice initiated, 159; historical value, 165-168.
Survey records, 216-219.
Surveying, problems, 21-22, 51-56, 66, 88-89, 97, 210, 212, 222-223; costs, 149-154; activities in 1800, 206-219; execution of, 220, 223-225; contributions to mapping and history, 220, 228-229.
Surveyors, 105, 123-128, 133-134, 161, 198, 207-208, 223-224; control measures, 51, 70-72; appointment of, 70; duties under proposed ordinance, 70-71, 76, 77, 78, 81; under warrant system, 70-71, 183; competence of, 72; duties under Ordinance of 1785, 94-97, 101-102, 104, 160; compensation, 101, 151-154; financial problems, 151-153; at land auction as advisors, 156; duties under Land Act of 1796, 196, 198-199; deputies under Putnam, 207-208.
Surveyors' chains, see Chains, surveyors'.

Surveyors General, 65, 66, 101n, 207, 215, 216, 221; of New England and other states, 68n-69n; office proposed, 186; under Land Act of 1796, 198, 199; appointments to, 200, 215, 216; of New York, 200; importance of, 207; Northwest of the Ohio, 213, 216. See also Mansfield, Jared; Putnam, Rufus, and Tiffin, Edward.
Surveyors' helpers, 152, 208, 209.
Symmes, John Cleves, 123n, 186, 195n; purchased land tract, 126, 170; proprietor, Miami Purchase, 180, 181n, 182, 183; failure, 184; and Indians, 187; failure to fulfill contract, 191.
Symmes Purchase, see Miami Purchase.

TATE, William, 134n.
Tatom, Absalom, 124n, 133n.
Tennessee, boundaries, 26, 28.
Tennessee River, 25n, 29n.
Territory North West of the River Ohio, 227; Virginians' expeditions into, 6; Indian threat in, 187.
Thompon, Charles, (text has Thompson), on state cessions, 15n.
Tiffin, Edward, first Commissioner, General Land Office, 206; Surveyor General, 216; closure method of, 216.
Townships, 41, 42, 59, 90, 92-96, 195, 214; in New England compared to hundreds, 45; survey of modern, 53; boundaries, 55, 146-149; term introduced, 56n; survey and sale contemplated, 87; size of, under Ordinance of 1785, 96, 104; numbering of, 97-98, 99; plats of, 164, 196; in Ohio Company purchase, 178-179; numbering of, in Miami Purchase, 182; numbering of, under Land Act of 1796, 197, 210; size under Land Act of 1796, 194; subdivision of, 205; appearance in Ordinance of 1785, 220; New England origin, 221.
Trading posts, 14n.
Trails, slighted in early survey notes, 166; historical data on, 166, 229.
Transit instrument, 121.
Treasurer of the United States, and administration of public lands, 69.
Treaties, Fort Finney (1786), 15n, 130n, 190; Fort Greenville (1795), 15, 188, 190; Fort Harmar (1789), 15n, 177n; Fort McIntosh (1785), 14, 82, 114, 116, 129, 190; Fort Stanwix (1784), 13-14; Jay's (1794), 188.
Trees, as survey markers, 73-74, 79, 81, 97, 104, 158-159, 166-168, 196, 202, 229; mentioned in Hutchins' report, 130; bearing trees, 158, 159n, 216; "line" trees, 158, 159n, 216; corner trees, 196, 216; early precedent for marking, 224.
Tupper, Anselm, 161n, 179n.
Tupper, Benjamin, 124, 125, 133n, 139n, 160-161.
"Tuscarawas," 135, 202n; Delaware salient at, 114, 115; Indian depredation at, 129.
Tuscarawas River, 190n, 203.

UNITED STATES DEPARTMENT OF THE INTERIOR, as land agency, 207n, 209n.
United States in Congress Assembled, see Continental Congress.
United States Military Reserve, 170, 202; surveying of, 202, 203; subdivision, 207n, 208n.
United States Treasury Department, and administration of public lands, 206, 217n. See also Secretary of the Treasury.

VAN ALEN (text has Van Allen), John E., 194.
Vegetation, along first survey line, 130-131.
Vermont, survey in, 90.
Veterans' state, see Reserves, military.
Vincennes, Indiana, 6, 10, 35, 142.
Virginia, western land cessions, 3, 19, 20, 32-33; western land claim, 5, 6, 7, 28; land act of 1779, 7, 41, 44n, 68, 72; lands reserved from cession, 8-9, 10, 11; boundary, 28; divided into hundreds, 43-44; land laws as model for federal system, 58, 68, 73, 76; boundary survey, 103, 119; survey agent, 127, 133; pioneer settlements, 133.
Virginia Military District, 8, 10, 11, 42, 95, 199, 205n.

WABASH RIVER, 33, 35, 191n.
Warrant system, see Land warrant system.
Washington, George, 25, 93n, 123n, 200; and Newburgh Petition, 22; on state boundaries, 29, 35; plan for western states, 29-31; as surveyor, 70-71, 73n, 74n; opposed rectilinear grid, 89; quoted on advantages of Ohio country, 103; on advantages of Seven Ranges, 107, 117-118; approved Land Act of 1796, 192; advocated indiscriminate surveys, 229.
"Washington," state of, proposed, 22, 105, 146.
Watercourses, names for, in early survey notes, 167-168; shown on Mansfield's map, 219.
Wayne, General Anthony, victory over Indians, 15, 188; and Indian boundary, 190n.
Wea Indians, 190n.
Webster, Pelatiah, 87n.
Wells, Charles, 111, 139.
Wentworth, Benning, 90n.
Western Reserve, see Connecticut Western Reserve.
Westmoreland County, Pennsylvania, 6n.
Wheeling, Virginia, 111.
Whipple, Levi, 207n.
Whitney, Charles, land speculation of, 192.
Williamson, Hugh, 72, 85, 94; member committee on western lands, 3; favored committee report, 4; on committee on Indian affairs, 13; and state cessions, 15n; as founder rectangular survey system, 38, 66, 225, 226; quoted on uniform grid, 40-41; aware of convergency problem, 51; and transmission of Dutch system, 63; and bill of 1792, 186, 187, 193, 194, 226; argued in favor of rectangular survey, 230. See also Jefferson-Williamson plan.
Wilson, William, 129n.
Wisconsin, as wilderness in 1848, 35.
Wolcott, Oliver, 227; commissioner at Fort Stanwix, 14n; and administration of public lands, 206, 207, 208, 209, 214, 217n.
Wood, James, 201n.
Worthington, Thomas, as surveyor and land office official, 207n, 208n.
Wyandott Indians, 114-115, 116, 129, 132; and Treaty of Fort McIntosh, 14; refused safe conduct to surveyors, 135; hostility, 138n, 187; land cessions, 188, 190.

YAZOO COMPANIES, 191.

ZANE, Ebenezer, 111, 200n.
Zane, Jonathan, 111n.
Zane's Trace, 111n; survey of, 200.

THE MANAGEMENT OF PUBLIC LANDS IN THE UNITED STATES

An Arno Press Collection

Abrams, Charles. **Revolution in Land.** 1939

Barnes, Will[iam] C[roft]. **Western Grazing Grounds and Forest Ranges.** 1913

Bogue, Margaret Beattie. **Patterns From the Sod.** 1959

Brayer, Herbert O. **Pueblo Indian Land Grants of the "Rio Abajo," New Mexico.** 1938

Calef, Wesley. **Private Grazing and Public Lands.** 1960

Carlson, Theodore L[eonard]. **The Illinois Military Tract.** 1951

Carstensen, Vernon. **Farms or Forests.** 1958

Chandler, Alfred N. **Land Title Origins.** 1945

Chapman, Berlin B. **Federal Management and Disposition of the Lands of Oklahoma Territory, 1866-1907** (Doctoral Dissertation, University of Wisconsin, 1931). 1979

Clawson, Marion. **The Western Range Livestock Industry.** 1950

Coles, Harry Lewis. **A History of the Administration of Federal Land Policies and Land Tenure in Louisiana, 1803-1860** (Doctoral Thesis, Vanderbilt University, 1949). 1979

Cotroneo, Ross Ralph. **The History of the Northern Pacific Land Grant, 1900-1952** (Doctoral Dissertation, University of Idaho, 1966). 1979

Diller, Robert. **Farm Ownership, Tenancy, and Land Use in a Nebraska Community.** 1941

Ernst, Joseph W. **With Compass and Chain** (Doctoral Dissertation, Columbia University, 1958). 1979

Gates, Paul Wallace. **Fifty Million Acres.** 1954

Gates, Paul Wallace, editor. **The Fruits of Land Speculation.** 1979

Gates, Paul Wallace and Robert W. Swenson. **History of Public Land Law Development.** 1968

Gates, Paul Wallace, editor. **Public Land Policies.** 1979

Gates, Paul Wallace, editor. **The Rape of Indian Lands.** 1979

Greever, William S. **Arid Domain.** 1954

[Harrison, Fairfax]. **Virginia Land Grants.** 1925

Hutchinson, William T. **The Bounty Lands of the American Revolution in Ohio** (Doctoral Dissertation, University of Chicago, 1927). 1979

Kilfoil, Jack F. **C. C. Trowbridge** (Doctoral Dissertation, Claremont Graduate School, 1969). 1979

Lampen, Dorothy. **Economic and Social Aspects of Federal Reclamation.** 1930

Lang, Aldon Socrates. **Financial History of the Public Lands in Texas.** 1932

La Potin, Armand Shelby. **The Minisink Patent** (Doctoral Thesis, University of Wisconsin, 1974). 1979

Lee, Lawrence Bacon. **Kansas and the Homestead Act, 1862-1905** (Doctoral Dissertation, University of Chicago, 1957). 1979

McKitrick, Reuben. **The Public Land System of Texas, 1823-1910.** 1918

Minneman, Paul George **Large Land Holdings in Ohio and Their Operation** (Doctoral Dissertation, The Ohio State University, 1929). 1979

Mullan, John. **Reports to the Hon. George Stoneman. Governor of California, on Certain Claims of the State of California Against the United States.** 1886

Nash, Gerald D. **State Government and Economic Development.** 1964

Odell, Marcia Larson. **Divide and Conquer** (Doctoral Dissertation, Cornell University, 1975). 1979

Okada, Yasuo. **Public Lands and Pioneer Farmers, Gage County, Nebraska, 1850-1900.** 1971

Ottoson, Howard W., et al. **Land and People in the Northern Plains Transition Area.** 1966

Pattison, William D. **Beginnings of the American Rectangular Land Survey Systems, 1704-1800.** 1970

Rakestraw, Lawrence. **A History of Forest Conservation in the Pacific Northwest, 1891-1913** (Doctoral Dissertation, University of Washington, 1955). 1979

Robinson, W[illiam] W[ilcox]. **Land in California.** 1948

Rollins, George W[atson]. **The Struggle of the Cattleman, Sheepman, and Settler for Control of Lands in Wyoming, 1867-1910** (Doctoral Thesis, University of Utah, 1951). 1979

Strausberg, Stephen. **Federal Stewardship on the Frontier** (Doctoral Dissertation, Cornell University, 1970). 1979

Teele, Ray P[almer]. **The Economics of Land Reclamation in the United States.** 1927

Van Name, Willard G. **Vanishing Forest Reserves.** 1929

Wilkinson, Norman B. **Land Policy and Speculation in Pennsylvania, 1779-1800** (Doctoral Dissertation, University of Pennsylvania, 1958). 1979

[Zonlight], Margaret Aseman Cooper. **Land, Water, and Settlement in Kern County, California, 1850-1890** (M. A. Thesis, University of California at Berkeley, 1954). 1979

ST. MARY'S COLLEGE OF MARYLAND
ST. MARY'S CITY, MARYLAND 20686

091905